Principles of
Ecotoxicology

Principles of Ecotoxicology

C. H. WALKER, S. P. HOPKIN, R. M. SIBLY AND D. B. PEAKALL

*School of Animal and Microbial Sciences,
University of Reading, UK*

Taylor & Francis
Publishers since 1798

UK   Taylor & Francis Ltd, 1 Gunpowder Square, London EC4A 3DE
USA   Taylor & Francis Inc., 1900 Frost Road, Suite 101, Bristol, PA 19007

British Library Cataloguing in Publication Data

A catalogue record for this book is available from the British Library.

ISBN 0-7484-0220-9
ISBN 0-7484-0221-7

Library of Congress Cataloging Publication Data are available

Cover design by Youngs Design in Production

Typset in Times 10/12pt by Santype International Ltd., Salisbury, Wilts

Printed in Great Britain by T. J. International Ltd

Contents

Preface

The origins of this book lie in the MSc course 'Ecotoxicology of Natural Populations' which was first taught at Reading in 1991. In recent years ecotoxicology has emerged as a distinct subject of interdisciplinary character. The structure of the course reflects this, and it is taught by people of widely differing backgrounds ranging from chemistry and biochemistry through to population genetics and ecology. Putting the different disciplines together in an integrated way was something of a challenge.

Experience of teaching the course persuaded the authors of the need for a textbook which would deal with the basic principles of such a wide-ranging subject. The intention has been to approach ecotoxicology in a broad interdisciplinary way, cutting across traditional subject boundaries. However, the nature of the text is bound to reflect the experience and interests of the authors, which will now be briefly reviewed.

Steve Hopkin is a Zoologist who has worked on electron microscopy and X-ray analysis for his PhD, and later investigated the effects of metals on soil ecology at the University of Bristol. Since coming to Reading his teaching and research has focused on the role of essential and non-essential metals in the biology of soil invertebrates.

David Peakall originally graduated as a chemist, and commenced his research as a physical chemist. Over a period he moved into biochemistry and finally into environmental toxicology. The last move was in keeping with his long-standing interest and active involvement in ornithology. During the last 15 years of his scientific career, he was chief of the Wildlife Toxicology division of the Canadian Wildlife Service, where he had a major involvement in studies of the Great Lakes.

Richard Sibly applied a degree in mathematics first in animal behaviour and then more widely in population biology. He has particular interests in life-history evolution and trade-offs, and in how these may be affected by environmental pollutants.

Colin Walker originally qualified as an agricultural chemist, and was responsible for chemical and biochemical studies on environmental pollutants at Monks Wood Experimental Station during the mid 1960s when the major concern was about the effects of organochlorine insecticides. He subsequently moved to the University of Reading where he has developed teaching and research into the molecular basis of toxicity, with particular reference to eco-toxicology.

Introduction

The term 'ecotoxicology' was introduced by Truhaut in 1969 and was derived from the words 'ecology' and 'toxicology'. The introduction of this term reflected a growing concern about the effects of environmental chemicals upon species other than man. It identified an area of study concerned with the harmful effects of chemicals (toxicology) within the context of ecology. Up to this time the subject of environmental toxicology had been principally concerned with the harmful effects of environmental chemicals upon man, e.g. the effects of smoke upon urban communities. However, environmental toxicology, in its widest sense, encompasses the effects of chemicals upon ecosystems as well as upon man. Thus, ecotoxicology is a discipline within the wider field of environmental toxicology. In the present text it is defined as 'the study of harmful effects of chemicals upon ecosystems'.

Despite the definition given above, much early work answering to the description of ecotoxicology had little 'ecology' or 'toxicology' about it. It was concerned with the detection and determination of chemicals in samples of animals and plants. Seldom could the analytical results be related to effects upon individual organisms, let alone effects upon populations or communities. Analytical techniques such as gas chromatography, thin-layer chromatography and atomic absorption facilitated the detection of very low concentrations of chemicals in biota; establishing the biological significance of these residues was a more difficult matter! One of the main themes of the present text is the problem of progressing from the measurement of concentrations of environmental chemicals to establishing their effects at the levels of the individual, the population and the community.

New disciplines frequently present problems of terminology, and ecotoxicology is no exception to this trend. Several important terms in ecotoxicology are used inconsistently in the literature. Their use in the present text will now be explained. Both 'pollutants' and 'environmental contaminants' are regarded as chemicals which exist at levels judged to be above those that would normally occur in any particular component of the environment. This immediately raises the question: what is to be considered normal? With most man-made organic chemicals, such as pesticides, the situation is simple – any detectable level is abnormal since the compounds did not exist in the environment until released by man. On the other hand chemicals such as heavy metals, sulphur dioxide, nitrogen oxides, polycyclic aromatic hydrocarbons (PAHs), and methyl mercury are naturally occurring and were present in the environment before the appearance of man. In the nature of things there is a variation in the concentration of these chemicals from place to place and from time to time. This makes it difficult to judge their normal ranges.

The distinction sometimes made between 'pollutants' and 'contaminants' raises further difficulties. The term 'pollutant' is taken to indicate that the chemical it describes is causing actual environmental harm, whereas the term 'contaminant' implies the chemical is not harmful. The difficulties with this distinction are threefold. First, there is the general toxicological principle that toxicity is related to dose (Chapter 5). Thus, a compound may answer to the description of pollutant in one situation but not another – a problem mentioned earlier. Second, there is no general agreement about what constitutes environmental harm or 'damage'. Some scientists would regard deleterious biochemical changes in an individual organism as harmful – others would reserve the term for declines in populations. Third, the effects of measured levels of chemicals in living organisms – or in their environment – are seldom known, yet the term 'pollutant' is frequently applied to them. Judgement of this issue is made more difficult by the possibility that there may be potentiation of toxicity when organisms are exposed to mixtures of environmental chemicals. To minimize these problems of terminology, the term 'pollutant' will be applied to environmental chemicals which exceed normal background levels and have the potential to cause harm. It would be attractive to reserve the term for particular chemicals in situations where they have been shown to cause harm, but because of the measurement problems referred to above this usage would be too restrictive. 'Harm' will be taken to include biochemical or physiological changes which adversely affect individual organisms' birth, growth or mortality rates. Such changes would necessarily produce population declines were it not that other processes (e.g. density-dependence) may compensate (Chapter 12).

Whether or not a contaminant is a pollutant therefore depends on its level in the environment, the organism being considered, and on whether or not the organism is harmed. Thus, a compound may answer to the description of 'pollutant' for one organism but not for another. Because of the problems in demonstrating harmful effects in the field, the terms 'pollutant' and 'contaminant' will, to a large extent, be used synonymously because it can seldom be said that contaminants have no potential to cause environmental harm in any situation. The term 'environmental chemical' will be used to describe any chemical that occurs in the environment without making any judgement as to whether it should be regarded as a pollutant or a contaminant.

Another word that has been used inconsistently in the literature is the term 'biomarker'. Here biomarkers are defined as *biological responses to environmental chemicals which give a measure of exposure and sometimes, also, of toxic effect*. Biomarker responses may be at the molecular, cellular or 'whole-organism' level. Some workers would regard population responses (changes in number or gene frequency) as biomarkers. However, since the latter tend to be much longer term than the former, it may be unwise to use the same term for both. In the present text the term biomarker will be restricted to biological responses at the level of the whole organism or below. An important thing to emphasize about biomarkers is that they represent measurements of effects, which can be related to the presence of particular levels of environmental chemical; they provide a means of interpreting environmental levels of pollutants in biological terms.

Finally, the organic pollutants to be considered here are examples of 'xenobiotics' ('foreign compounds'). They play no part in the normal biochemistry of living organisms. The concept of 'xenobiotics' will be discussed further in Chapter 5.

An exciting feature of ecotoxicology is that it represents a 'molecules to ecosystems' approach which relates to the 'genes-to-physiologies' approach originally identified by Clarke (1975) and extensively developed in North America in the 1980s (see e.g. Feder *et al.* 1987). Moreover it analyses 'experimental' manipulations on the largest of scales (though the 'experiments' were not designed as such). Thus heavy-metal pollution, acid rain, and application of pesticides have affected whole ecosystems, sometimes with dramatic consequences for the populations within them. In ecotoxicology the ecosystem response is studied at all levels. Initially (see figure), the molcular structures of pollutants, their properties and environmental fate are considered (Part One of book).

Ecophysiologists generally analyze the impact of pollutants on an organism's growth, birth and death rates; indeed, as explained above, pollutants can adversely affect these 'vital rates'. This makes it desirable to understand how adverse effects on vital rates have implications for populations (Chapters 12 and 13). Thus, the relationship between the vital rates and 'population growth rate' is described in detail in Chapter 12. Consequently it is, in principle, possible to evaluate pollutants quantitatively in terms of their population effects. This emphasis on vital rates as crucial intervening variables, linking physiological effects to population effects, is a particular feature of this book. The approach is continued in Chapter 13 to consider whether and how fast resistant genes increase in populations. The rate at which resistant genes increase is measured by the 'population growth rate' of the 'population' of resistant genes. The 'population growth rate' of resistant genes is a measure of their Darwinian fitness. Although this is not the conventional population-genetic measure of fitness, it is particularly useful in ecotoxicology because it (alone) shows explicitly how the fitness of resistant genes depends on the effects those genes have on their carriers. To summarize, the approach taken in this book allows linkage to be made between the different levels of organization shown in the figure, from molecules to physiologies to populations, right through to ecosystems. This is the underlying basis for the biomarker strategy which seeks to measure sequences of responses to pollutants from the molecular level to the level of ecosystems (Chapters 10 and 15). The employment of biomarkers in biomonitoring is described in Chapter 11. These three chapters are placed at the end of their respective sections of the book. They represent the practical realization of theoretical aspects described in earlier chapters.

The text is divided into three parts:

Part One describes major classes of organic and inorganic pollutants, their entry into the environment and their movement, storage and transformation within the environment. Thus, it bears a certain resemblance to toxicokinetics in 'classical' toxicology, which is concerned with the uptake, distribution, metabolism and excretion of xenobiotics by living organisms (Chapter 5). The difference is one of complexity. Ecotoxicology deals with movements of pollutants in air, water, soils and sediments, and through food chains, with chemical transformation and biotransformation.

Part Two deals with the effects of pollutants upon living organisms, thus resembling toxicodynamics in classical toxicology. The difference is again one of complexity. Whereas toxicodynamics focuses upon interactions between xenobiotics and their sites of action, ecotoxicology is concerned with a wide range of effects upon individual organisms at differing organizational levels (molecular, cellular and whole animal). Toxicity data obtained in the labor-

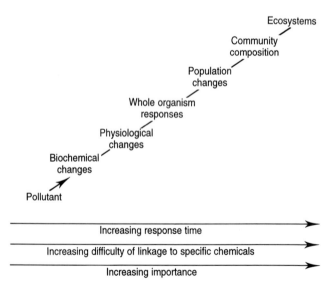

Schematic relationship of linkages between responses at different organization levels.

atory are used for the purposes of risk assessment. Effects of pollutants are studied in the laboratory, an approach that can lead to the development of biomarker assays (Chapter 10). The use of biomarker assays in biomonitoring is discussed in Chapter 11, which also considers some effects at the population level, thereby looking ahead to the final part of the text.

Part Three addresses questions which are of the greatest interest to ecologists. What effects do pollutants have at the level of population, community, and whole ecosystem? This takes the discussion into the disciplines of population biology and population genetics. Whereas classical toxicology is concerned with chemical toxicity to individuals, ecotoxicologists are particularly interested in effects at the level of population community and whole ecosystem. Effects at the population level may be changes in numbers of individuals (Chapter 12), changes in gene frequency (as in resistance) (Chapter 13) or changes in ecosystem function (e.g. soil nitrification) (Chapter 14). They may be due to sublethal effects (e.g. on physiology or behaviour) rather than lethal toxicity. Sometimes they may be indirect (e.g. the decline in a predator due to direct chemical toxicity may lead to an increase in numbers of its prey). It is often very difficult to establish effects of pollutants on natural populations. However, the development of appropriate biomarker assays can help to resolve this problem.

The third part of this book illustrates the truly interdisciplinary character of ecotoxicology. The study of the harmful effects of chemicals upon ecosystems draws on the knowledge and skills of ecologists, physiologists, biochemists, toxicologists, chemists, meteorologists, soil scientists and others. It is nevetheless a discipline with its own distinct character. Apart from the important applied aspects which address current public concerns, it has firm roots in basic science. Chemical warfare is nearly as old as life itself, and the evolution of detoxication mechanisms by animals to avoid the toxic effects of xenobiotics produced by plants is parallelled by the recent development of resistance by pests to pesticides made by humans.

Acknowledgments

Many people have contributed to this book in all sorts of ways. While we cannot acknowledge them all, we would particularly like to mention our MSc students, who have contributed much in discussion and feed back, and Amanda Callaghan, Peter Dyte, Andy Hart, Graham Holloway, Alan McCaffery, Mark Macnair, Ian Newton, Demetris Savva, Ken Simkiss, Nick Sotherton, and George Warner. Last, but not least, Gill Bogue and Val Walker who have given invaluable secretarial support.

Figure Permissions

The following figures are reproduced with the kind permission of the publishers.

Figure 1.1 © (1989) Chapman & Hall
Figure 1.2 © (1989) Chapman & Hall
Figure 2.1 © (1975) Macmillan
Figure 2.2 © (1975) Macmillan
Figure 3.2 © (1976) Prentice Hall
Figure 3.3 © (1992) Prentice Hall
Figure 3.4 © (1992) John Wiley & Sons Ltd
Figure 4.1 © (1989) Chapman & Hall
Figure 4.3 © (1994) John Wiley & Sons Ltd
Figure 4.4 © (1993) Elsevier Science Ltd
Figure 4.5 © (1994) Williams & Wilkins
Figure 4.6 © (1993) National Research Council of Canada
Figure 4.7 © (1980) Elsevier Science Ltd
Figure 4.8 © (1989) Elsevier Science Publishers
Figure 4.9 © (1994) Elsevier Science Ltd
Figure 5.1 © (1993) Appleton & Lange
Figure 5.6A © (1987) Academic Press
Figure 5.6B © (1989) Elsevier Science Inc
Figure 5.8 © Academic Press
Figure 6.2 © (1993) Blackwell Science Ltd
Figure 6.4 © (1994) Elsevier Science Ltd
Figure 6.7 © (1993) Academic Press
Figure 6.10 © (1993) Blackwell Science Ltd
Figure 7.4 © (1990) Marcel Dekker
Figure 8.1 © (1992) John Wiley & Sons Ltd
Figure 8.2 © (1992) John Wiley & Sons Ltd

Figure 8.3 © (1990) Blackwell Science Ltd
Figure 8.4 © (1990) Blackwell Science Ltd
Figure 8.5 © (1989) Chapman & Hall
Figure 8.6 © (1984) Zoological Society of London
Figure 8.7 © (in press) Lewis Publishers
Figure 8.8 © (1994) Springer-Verlag
Figure 8.9 © (1994) Springer-Verlag
Figure 8.11 © (1992) Elsevier Science B.V.
Figure 8.12 © (1989) Linnean Society of London
Figure 8.13 © (1989) Linnean Society of London
Figure 8.14 © (1989) Linnean Society of London
Figure 9.1 © (1989) Elsevier Applied Science
Figure 10.1 © (1993) Springer-Verlag
Figure 11.1 © (1993) Blackwell Science Ltd
Figure 11.2 © (1987) Elsevier Science Ltd
Figure 11.3 © (1990) Entomological Society of America
Figure 11.4 © (1994) Williams & Wilkins
Figure 11.5 © (1993) Blackwell Science Ltd
Figure 11.6 © (1993) Blackwell Science Ltd
Figure 11.7 © (1991) Elsevier Science Ltd
Figure 11.8 © (1994) Marine Biological Association of the UK
Figure 12.3 © (1981) National Research Council of Canada
Figure 12.4 © (in press) Lewis Publishers
Figure 12.5 © (1954) University of Chicago Press
Figure 12.9 © (1993) National Research Council of Canada
Figure 12.10 © (1986) Collins
Figure 12.11 © (1986) Collins
Figure 12.12 © (1986) Collins
Figure 12.13 © (1986) Collins
Figure 12.14 © (1986) Collins
Figure 12.16 © (1986) Poyser
Figure 12.17 © (1993) Poyser
Figure 12.18 © (1993) Poyser
Figure 12.19 © (1988) Elsevier Science Ltd
Figure 12.20 © (1992) Blackwell Science Ltd
Figure 12.21 © (1992) HMSO
Figure 12.22 © (1992) HMSO
Figure 13.5 © (1990) Blackwell Science Ltd
Figure 13.6 © (1986) Linnean Society of London
Figure 13.9 © (1973) Oxford University Press
Figure 13.10 © (1987) Elsevier Trends Journals
Figure 14.4B © (1983) National Research Council of Canada

Pollutants and their fate in ecosystems

1

Major Classes of Pollutant

Many different chemicals are regarded as pollutants, ranging from simple inorganic ions to complex organic molecules. In the present chapter, representatives will be identified of all the major classes of pollutant, and their properties and occurrence will be briefly reviewed. These pollutants will be used as examples throughout the text. Their fate in the living environment will be the subject of the remainder of Part One. Their effects upon individuals and ecosystems will be considered in Parts Two and Three respectively.

1.1 Inorganic Ions

1.1.1 Metals

A metal is defined by chemists as being an element which has a characteristic lustrous appearance, is a good conductor of electricity, and generally enters chemical reactions as positive ions or cations.

Although metals are usually considered as pollutants, it is important to recognize that they are natural substances. With the exception of radioisotopes produced in man-made nuclear reactions (bombs and reactors), all metals have been present on the earth since its formation. There are a few examples of localized metal pollution resulting from natural weathering of ore bodies (e.g. Hågvar and Abrahamsen, 1990). However, in most cases, metals become pollutants where human activity, mainly through mining and smelting, releases them from the rocks in which they were deposited during volcanic activity or subsequent erosion and relocates them into situations where they can cause environmental damage.

The extent to which human activity contributes to global cycles of metals can be described by the *anthropogenic enrichment factor* (AEF) (Table 1.1).

3

Table 1.1 Anthropogenic enrichment factors (AEF) for total global annual emissions of cadmium, lead, zinc, manganese and mercury in the 1980s (all values 10^6 kg y^{-1})^a

Metal	Anthropogenic sources (A) (industry etc.)	Natural sources (volcanoes etc.)	Total (T)	AEF $= \dfrac{A}{T} \times 100$
Cadmium (Cd)	8	1	9	89%
Lead (Pb)	300	10	310	97%
Zinc (Zn)	130	50	180	72%
Manganese (Mn)	40	300	340	12%
Mercury (Hg)	100	50	150	66%

^{a} *Note:* From various sources.

From this table, it is clear that human activity is responsible for the majority of the global movement of cadmium, lead, zinc and mercury, but is relatively unimportant in the cycling of manganese. The high AEF for lead is due mostly to the widespread use and subsequent release of lead-based additives to petrol. For most radioactive isotopes, the AEF is 100%.

Elements that are considered to be metals are shown in Figure 1.1. The term *heavy metals* has been used extensively in the past to describe metals which are environmental pollutants. For a metal to be considered 'heavy' it must have a density relative to water of greater than five. However, the term

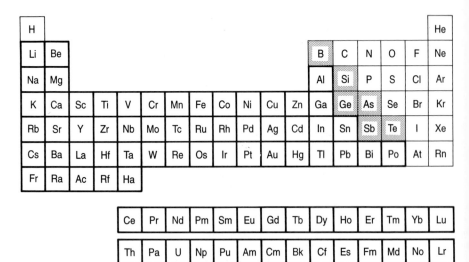

Figure 1.1 Periodic table of the elements. Those considered to be metals are surrounded by bold lines. Metalloids (with properties of metals and non-metals) are shaded. Reproduced from Hopkin (1989) with permission of Elsevier Applied Science.

'heavy metals' has been replaced in recent years by a classification scheme that considers their chemistry rather than relative density (Nieboer and Richardson, 1980; Table 1.2). This approach is more logical since there are some metals that are not 'heavy' which can be important environmental pollutants. Aluminium, for example, which is a metal, has a relative density of only 1.5. However, it is an extremely important pollutant in acidified lakes where it becomes soluble and is toxic to fauna. The gills of fish are particularly susceptible to aluminium poisoning. Aluminium has also been implicated in Alzheimer's disease in humans and may be deposited in the brain.

Metals are *non-biodegradable*. Unlike some organic pesticides, metals cannot be broken down into less-harmful components. Detoxification by organisms consists of 'hiding' active metal ions within a protein such as metallothionein (binding covalently to sulphur), or depositing them in an insoluble form in intracellular granules for long-term storage or excretion in the faeces (see Chapter 8).

Essential elements all have a 'window of essentiality' within which dietary concentrations in animals, or soil concentrations in plants, have to be maintained if the organism is to grow and reproduce normally (Figure 1.2). In addition to carbon, hydrogen, oxygen and nitrogen, all animals need the seven major mineral elements calcium, phosphorus, potassium, magnesium, sodium, chlorine and sulphur for ionic balance, and as integral parts of amino acids, nucleic acids and structural compounds. Thirteen other so-called 'trace elements' are definitely required, namely, iron, iodine, copper, manganese, zinc,

Table 1.2 Separation of some essential and non-essential metal ions of importance as pollutants into class A (oxygen-seeking), class B (sulphur- or nitrogen-seeking) and borderline elements based on the classification scheme of Nieboer and Richardson (1980)[a]

Class A	Borderline	Class B
Calcium	Zinc	Cadmium
Magnesium	Lead	Copper
Manganese	Iron	Mercury
Potassium	Chromium	Silver
Strontium	Cobalt	
Sodium	Nickel	
	Arsenic	
	Vanadium	

[a] *Note:* This distinction is important in determining rates of transport across cell membranes, and sites of intracellular storage in metal-binding proteins and metal-containing granules (e.g. section 8.2).

5

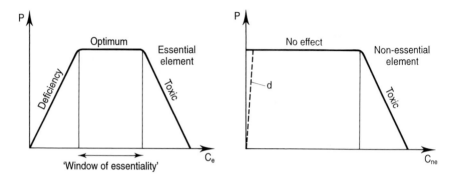

Figure 1.2 Relationships between performance (P) (growth, fecundity, survival) and concentrations of an essential (C_e) or non-essential (C_{ne}) element in the diet of animals. Possible deficiency effects at ultra-trace levels (d) of an apparently non-essential element may be discovered as the sensitivities of analytical techniques are improved. Reproduced from Hopkin (1989) with permission of Elsevier Applied Science.

cobalt, molybdenum, selenium, chromium, nickel, vanadium, silicon and arsenic. Zinc, for example, is an essential component of at least 150 enzymes, copper is essential for the normal function of cytochrome oxidase, and iron is part of haemoglobin, the oxygen-carrying pigment in red blood cells. Boron is required exclusively by plants. A few other elements, such as lithium, aluminium, fluorine and tin, may be essential at ultratrace levels. The window of essentiality for some elements is very narrow. Selenium for example was considered for a long time to be only a dangerous toxin until its role in the enzyme glutathione peroxidase was discovered (Jukes, 1985). The dose determines the poison.

Non-essential metals such as mercury or cadmium, in addition to being toxic above certain levels, may also affect organisms by inducing deficiencies of essential elements through competition at active sites in biologically important molecules (Table 1.3) (see Chapter 7). Such *antagonism* also occurs between essential elements. A concentration of only 5 μg Mo g^{-1} in the diet of cattle is sufficient to reduce copper uptake by 75%, which often leads to symptoms of copper deficiency.

1.1.2 *Anions*

There are some inorganic pollutants which are not particularly toxic, but which cause environmental problems because they are used in such large quantities. These include anions such as nitrates and phosphates.

Nitrate fertilizers are used extensively in agriculture. During the growth period of crops, most of the fertilizer applied is absorbed by plant roots. However, when growth ceases, nitrate released during the decomposition of

Table 1.3 Level of activity of carbonic anhydrase expressed relative to that of zinc, of different metals substituted in the protein[a]

Metal	Normal activity (hydration of CO_2) (%)
Zinc	100
Cobalt	56
Nickel	5
Cadmium	4
Manganese	4
Copper	1
Mercury	0.05

[a] *Note:* From Coleman (1967).

dead plant material passes down through the soil and may enrich adjacent water courses. The increase in available nitrogen may cause blooms in algal populations. This effect is called *eutrophication* and eventually leads to oxygen starvation as microorganisms break down the dead algal tissues.

The safe limit for nitrates in drinking water in the UK has been set at 100 parts per million. A human health problem may arise if young babies ingest bottled milk made up with nitrate-contaminated water. During their first few months of life, human infants have an anaerobic stomach. The nitrates are converted to nitrites in this oxygen-poor environment. The nitrites bind to haemoglobin, reduce its capacity to carry oxygen, and the infant may develop 'blue baby syndrome' or *methaemoglobinaemia*. The problem does not arise with breast-fed babies (definitely a case of 'breast is best'!). In regions of intensive agriculture, the 100 ppm level is exceeded in water extracted from rivers, or bore holes where nitrates have leached down to aquifers. The problem can be solved by removing the nitrate chemically at the water treatment works, or diluting the contaminated water with water from a relatively nitrate-free source. The long-term solution is of course to reduce nitrate usage, and this is being done in so-called 'exclusion zones' around sources of water for human consumption.

Similar problems of eutrophication can also arise with phosphates used as fertilizers. However, there is an additional source: washing powders. These have been made less resistant to breakdown in recent years due to cooperation between soap manufacturers and water-treatment companies. In the 1950s and 1960s, it was common to see a huge build up of foam below weirs and waterfalls downstream of the outfalls of sewage treatment works.

1.2 Organic Pollutants

The great majority of compounds that contain carbon are described as 'organic', the few exceptions being simple molecules such as CO_2 and CO. Carbon has the ability to enter into the formation of a bewildering diversity of

7

complex organic compounds, many of which provide the basic fabric of living organisms. The reason for this is the tendency of carbon atoms to form stable bonds with one another, thereby creating rings and extended chains. Carbon can also form stable bonds with hydrogen, oxygen and nitrogen atoms.

Molecules built of carbon alone (e.g. graphite and diamond), or of carbon and hydrogen (hydrocarbons), have very little polarity and consequently low water solubility. Polar molecules have electrical charge associated with them; non-polar molecules have little or none. Molecules with a strong charge are described as highly polar; molecules of low charge have low polarity. Polar compounds tend to be water soluble because the charges on them are attracted to opposite charges on water molecules. For example, a positive charge on an organic molecule will be attached to a negative charge on a water molecule. Carbon compounds tend to be more polar and more chemically reactive when they contain functional groups such as $-OH$, $-CH=O$ and $-NO_2$. In these examples, the oxygen atom attracts electrons away from neighbouring carbon atoms, thereby creating a charge imbalance on the molecule. Molecules of high polarity tend to enter into chemical and biochemical reactions more readily than do molecules of low polarity.

The behaviour of organic compounds is dependent upon their molecular structure – molecular size, molecular shape and the presence of functional groups being important determinants of metabolic fate and toxicity. Thus, it is important to know the formulae of pollutants in order to understand or predict what happens to them in the living environment. The principles operating here are illustrated by examples given in Chapters 5 and 7. Readers with a limited knowledge of chemistry are referred to the text of Manahan (1994), which contains two useful concise chapters on basic principles.

The pollutants that will be described here are predominantly man-made ('anthropogenic') compounds which have appeared in the natural environment only during the last century. This is only a very short time in evolutionary terms, and there has been only limited opportunity for the evolution of protective mechanisms against their toxic effects (e.g. detoxication by enzymes). In this respect they differ from inorganic pollutants, and from those naturally occurring xenobiotics, which have substantial toxicity (e.g. nicotine, pyrethrins and rotenone are compounds produced by plants which are highly toxic to certain species of insect). Aromatic hydrocarbons represent a special case. They have been generated by the combustion of organic matter since the appearance of higher plants on earth (e.g. due to forest fires started by volcanic lava). Like heavy metals that are mined, their environmental levels increase substantially as a consequence of human activity (as with the combustion of coal or petrol to produce aromatic hydrocarbons).

1.2.1 *Hydrocarbons*

These are compounds composed of the elements carbon and hydrogen only. Some hydrocarbons of low molecular weight (e.g. methane, ethane and ethylene) exist as gases at normal temperature and pressure. However, the great majority of hydrocarbons are liquids or solids. They are of low polarity (i.e. electrical charge, see above), and consequently have low water solubility, but high solubility in oils and in most organic solvents. (They are not very soluble in polar organic solvents such as methanol or ethanol.)

Hydrocarbons are divisible into two classes: (1) alkanes, alkenes and alkynes, and (2) *aromatic* hydrocarbons (Figure 1.3). The distinguishing feature of aromatic hydrocarbons is the presence of one or more benzene rings in their structure. Benzene rings are six-membered carbon structures which are 'unsaturated' in the sense that not all available carbon valences are taken by linkage to hydrogen. In fact, benzene rings have delocalized electrons which can move freely over the entire ring system and do not remain in the immediate vicinity of any one atom. Other hydrocarbons do not have this feature. They vary greatly in molecular size, and may be fully saturated (e.g. hexane, octane) or unsaturated. Unsaturated hydrocarbons contain carbon–carbon (C–C) *double bonds* (e.g. ethylene) or carbon–carbon *triple bonds* (e.g. acetylene). Saturated hydrocarbons are referred to as alk*anes*, unsaturated hydrocarbons with a carbon–carbon double bond are alk*enes* and unsaturated hydrocarbons with a carbon–carbon triple bond are alk*ynes*. They may exist

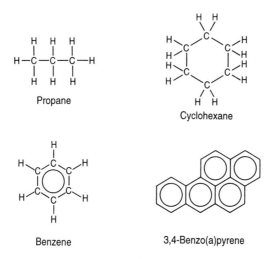

Propane

Cyclohexane

Benzene

3,4-Benzo(a)pyrene

Figure 1.3 Hydrocarbons. Composed of only hydrogen and carbon, these compounds have low polarity and thus low water solubility, but high solubility in oils and organic solvents. Propane and cyclohexane are examples of alkanes, benzene and benzo(a)pyrene of aromatic compounds. The latter contain six-membered carbon rings (benzene rings) with delocalized electrons. This is indicated by representing the benzene rings as a hexagon (the six-membered carbon frame) and a circle (the cloud of delocalized electrons) situated within it. Aromatic hydrocarbons undergo certain characteristic biotransformations influenced by the delocalized electrons (section 5.1.5).

as single chains, branched chains or rings (Figure 1.3). The properties of these two groups of hydrocarbons will now be considered separately.

The properties of non-aromatic hydrocarbons depend upon molecular weight and degree of unsaturation. Alkanes are essentially stable and unreactive, and have the general formula C_nH_{2n+2}. The first four members of the series exist as gases ($<n = 4$). Where $n = 5$–17, they are liquids at normal temperature and pressure. Where $n = 18$ or more they are solids. Alkenes and alkynes are more chemically reactive because they contain carbon–carbon double or triple bonds. As with alkanes, the lower members of the series are gases, the higher members liquids or solids.

Aromatic hydrocarbons exist as liquids or solids – none of them has a boiling point below 80°C at normal atmospheric pressure. They are more reactive then alkanes, being susceptible to chemical and biochemical transformation. There are many polycyclic aromatic hydrocarbons (PAHs) which are planar (i.e. flat) molecules consisting of three or more six-membered (benzene) rings directly linked together.

The major sources of hydrocarbons are deposits of petroleum and natural gas in the upper strata of the earth's crust. These fossil fuels originate from the remains of plants and animals of earlier geological times (notably the carboniferous period). While non-aromatic hydrocarbons predominate in these deposits, crude oils also contain significant amounts of PAHs. PAHs are also formed as a consequence of the incomplete combustion of organic materials. Thus they are generated when coal, oil or petrol are burned, when trees or houses burn, and when people smoke cigarettes. Major sources of hydrocarbon pollution are spillage of crude oils (e.g. oil tanker disasters) and the combustion of fossil fuels (notably the use of brown coal in parts of eastern Europe).

1.2.2 Polychlorinated Biphenyls (PCBs)

These are commercial mixtures of related compounds (congeners) which are useful for their physical properties. They are stable, unreactive viscous liquids, of low volatility, which have been used as hydraulic fluids, coolant–insulation fluids in transformers, and plasticizers in paints. There are altogether 209 possible PCB congeners, and some 120 of these are present in commercial products such as Aroclor 1254, Aroclor 1260 and Clophen A60. The last two digits in the numbers refer to the percentage chlorine in the PCB mixture. In either case, the larger the number the greater the proportion of higher chlorinated PCBs in the mixture.

PCB mixtures have very low solubility in water but high solubility in oils and organic solvents of low polarity. The water solubilities of Aroclor 1254 and Aroclor 1260, are only 21 μg l^{-1} and 2.7 μg l^{-1} respectively.

The individual congeners of PCB vary in their stereochemistry, depending on the positions of substitution of chlorine atoms. Where there is no substitution in the ortho positions, the two benzene rings tend to remain in the same

Major classes of pollutant

plane (coplanar PCBs), 3,3',4,4'-tetrachloro-biphenyl is an example of a coplanar PCB (Figure 1.4). By contrast, substitution of two, three or four ortho-
positions with chlorine leads to the movement of the rings out of plane, due to
the interaction of adjacent chlorines in different rings (chlorine atoms are
bulky). The molecular conformation is not then a coplanar (flat) one, but tends
more to a 'globular' structure.

PCBs were once used for a number of purposes – as dielectric fluids, heat
transformer fluids, lubricants, vacuum pump fluids, as plasticizers (e.g. in
paints) and for making carbonless copy paper. In many countries the use of
PCBs is now banned or severely restricted.

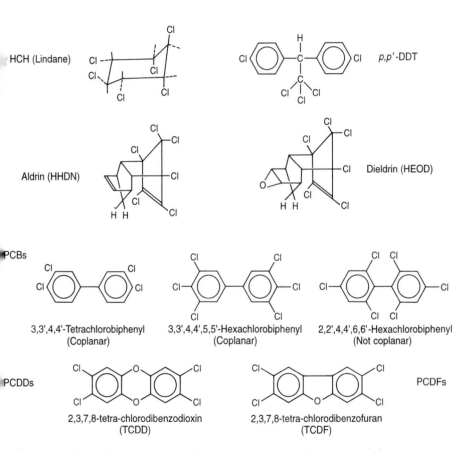

Figure 1.4 Organohalogen compounds are organic compounds containing *halogen* atoms
(the halogen elements are fluorine, chlorine, bromine and iodine). All the examples given
here are of organochlorine compounds, although it should be noted that organofluorine
compounds (e.g. chlorfluorocarbons) and organobromine compounds (e.g. polybrominated
biphenyls) are also environmental pollutants. The compounds shown here are stable solids of
low polarity and water solubility. They are not found in nature, and in many cases are only
slowly metabolized and consequently are persistent in living organisms (Chapter 5).

11

Major sources of pollution are (or have been) manufacturing wastes and the careless disposal or dumping of the liquids referred to above.

1.2.3 *Polychlorinated Dibenzodioxins (PCDDs)*

The best known member of this group of compounds is 2,3,7,8-tetra-chlorodibenzodioxin (Figure 1.4), usually referred to simply as 'dioxin'. This is a compound of extremely high toxicity to mammals (10–200 μg kg^{-1} in rats and mice). In structure these are 'flat' molecules, formed by the linking of two benzene rings by two oxygen bridges with varying substitutions of chlorine on the available ring positions. There are 75 possible congeners of PCDD. PCDDs are chemically stable compounds with very low water solubilities (less than 1 μg l^{-1} at 20°C), and limited solubility in most organic solvents, even though they have a lipophilic character.

PCDDs are not produced commercially, but are unwanted byproducts which are generated during the synthesis of other compounds. They are also formed during the combustion of PCBs (fires or chemical waste disposal), and by the interaction of chlorophenols during disposal of industrial wastes. In general, they are formed when chlorophenols interact. The problem of environment pollution by these compounds is best illustrated by reference to three well publicized examples involving 2,3,7,8-TCDD.

1. 2,3,7,8-TCDD ('dioxin') can occur as a contaminant of commercial preparations of herbicides such as 2,4D and 2,4,5T. There have been a number of investigations into pollution caused by these preparations, most notoriously the spraying by the US airforce of the Vietnamese jungle with the defoliant herbicide 'Agent Orange' during 1960–1969 (see section 1.2.10).
2. The release of PCDDs by combustion furnaces used to dispose of PCB wastes e.g. at the 'Rechem' plant in Scotland in the 1980s. PCDD residues were detected in soil and cattle in the surrounding area. The release of these compounds indicated incorrect operation of the furnace.
3. The release of PCDDs into the air from the Seveso Chemical Works in Northern Italy in 1976. The PCDDs were formed in a chemical cloud which contained trichlorophenols. Some people who had been exposed developed a skin condition, chloracne. There were, however, no fatalities or serious toxic effects attributable to PCDD.

PCDD residues have been detected very widely in the environment (especially in the aquatic environment), albeit at low concentrations e.g. in fish and fish-eating birds.

1.2.4 *Polychlorinated Dibenzofurans (PCDFs)*

These compounds are similar to PCDDs both in structure (Figure 1.4) and in origin. Once again there are many congeners, and the compounds arise as unwanted byproducts – they are not synthesized intentionally. They have not, however, received so much attention as PCDDs, and do not appear to have raised such serious environmental problems.

1.2.5 Polybrominated Biphenyls (PBBs)

Mixtures of polybrominated biphenyls have been marketed as fire retardants (e.g. 'Firemaster'). These mixtures bear a general resemblance to PCB mixtures, and are lipophilic, stable and unreactive. As with PCBs, some congeners are very persistent in living organisms and have long biological half-lives. In one incident in the USA, a PBB mixture was accidentally fed to cattle, leading to the appearance of substantial residues in meat products and humans in Wisconsin and neighbouring states.

1.2.6 Organochlorine Insecticides

This is a relatively large group of insecticides with considerable diversity of structure, properties and uses. Three major types will be mentioned here – DDT and related compounds, the chlorinated cyclodiene insecticides (e.g. aldrin and dieldrin) and hexachlorocyclohexanes (HCHs) such as Lindane (Figure 1.4).

Organochlorine insecticides are stable solids of limited vapour pressure, very low water solubility, and high lipophilicity. Some of them are highly persistent in their original form or as stable metabolites. All the examples given here are nerve poisons (see Chapter 7).

Commercial DDT contains 70–80% of the insecticidal isomer pp′DDT. Related insecticides include rhothane (DDD) and methoxychlor. The insecticidal properties of DDT were discovered by Paul Müller of the firm Ciba-Geigy in 1939. DDT was used to a small extent (mainly for vector control) during the Second World War, but came to be very widely used thereafter for the control of agricultural pests, vectors of disease (e.g. malarial mosquitoes), ectoparasites of farm animals, and insects in domestic and industrial premises. Because of its low solubility in water (<1 mg l^{-1}) DDT has been formulated as an emulsifiable concentrate, for application as a spray. (*Emulsifiable concentrates* are solutions of pesticides in organic liquids. When added to water, they form a creamy emulsion which can be sprayed on crops.) DDT has an LD$_{50}$ of 113–450 mg kg^{-1} in rats and is regarded as moderately toxic (Chapter 6).

Aldrin, dieldrin and heptachlor are all examples of cyclodiene insecticides. They resemble DDT in being stable solids, lipophilic and of low water solubility, but differ from it in their mode of action (Chapter 7). They are highly toxic to mammals (LD$_{50}$ 40–60 mg kg^{-1}). They came into use in the 1950s for controlling a variety of insect pests, notably as seed dressings and soil insecticides.

Hexachlorocyclohexane (HCH) has been marketed as a crude mixture of isomers ('BHC') but more extensively as a refined product containing mainly the γ isomer, known as γHCH, γBHC or lindane (Figure 1.4). γHCH has similar properties to other organochlorine insecticides, but is somewhat more polar and water soluble (7 mg l^{-1}). Emulsifiable concentrates of HCH have been used for controlling agricultural pests or parasites of farm animals. It has also been used (and to some extent still is) as an insecticidal seed dressing (e.g. on cereal seed). HCH is only moderately toxic to rats (LD$_{50}$ 60–250 mg kg^{-1}).

1.2.7 *Organophosphorus Insecticides (OPs)*

During the Second World War, interest developed in organophosphorus compounds which act as nerve poisons (neurotoxins) because of their ability to inhibit the enzyme acetylcholinesterase (AChE) (Chapter 7). These compounds were produced for two main uses – as insecticides and as chemical warfare agents (nerve gases). They are organic esters of phosphorus acids (Figure 1.5). Today, a large number of organophosphorus compounds are marketed as insecticides, and nearly all of them correspond to the basic formula shown in Figure 1.5.

Most organophosphorus insecticides (OPs) are liquids of lipophilic character, and some volatility; a few are solids. They are, in general, less stable than

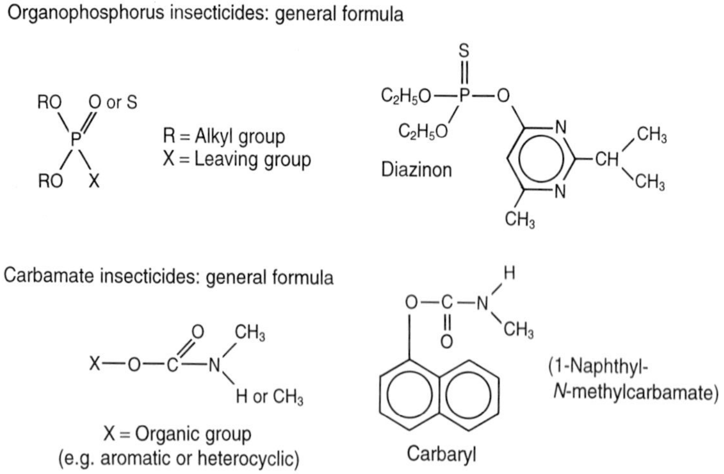

Figure 1.5 Organophosphorus and carbamate insecticides. These compounds are toxic to insects because they inhibit the enzyme acetylcholinesterase (Chapter 7). They vary in their polarity and water solubility. They are generally more reactive and less stable and persistent than the organochlorine insecticides. The leaving group of the organophosphorus compounds breaks away from the rest of the molecule when hydrolysis occurs (Chapter 5).

organochlorine insecticides and are more readily broken down by chemical or biochemical agencies. Thus, they tend to be relatively short-lived when free in the environment, and the environmental hazards that they present are largely, but not exclusively, associated with short-term (acute) toxicity. They are more polar and water-soluble than the main types of organochlorine insecticides. Their water solubility is highly variable, with some compounds (e.g. dimethoate) having appreciable solubility. The active forms of some OPs have sufficient water solubility to be effective systemic insecticides reaching high enough concentrations in the phloem of plants to poison sap feeding insects (cf. OCs and pyrethroids).

The formulation of organophosphorus compounds is important in determining the environmental hazards that they present. Many are formulated as emulsifiable concentrates for spraying. Others are incorporated into seed dressings or into granular formulations. Granular formulations are required for the most toxic OPs (e.g. disyston and phorate) because they are safer to handle than emulsifiable concentrates or certain other types of formulation. The insecticide is 'locked up' within the granule, and is only slowly released into the environment.

In many countries OPs are still applied to crops as sprays, granules, seed dressings and root dips. They are used to control ectoparasites of farm and domestic animals (commonly in sheep dipping), and sometimes also for controlling internal parasites (e.g. ox warble fly). Other uses include control of certain vertebrate pests (e.g. the bird Quelea in parts of Africa), locusts, stored product pests, especially beetles, insect vectors of disease, such as mosquitoes, and parasites of salmon at 'fish farms'.

1.2.8 Carbamate Insecticides

These are derivatives of carbamic acid which have been developed more recently than OCs and OPs (Figure 1.5). Like OPs, however, they act as inhibitors of acetylcholinesterase. Carbamates are frequently solids, sometimes liquids. They vary greatly in water solubility. Like OPs, they are readily degradable by chemical and biochemical agencies and do not usually raise problems of persistence. The main hazards that they present relate to short-term toxicity. Some of them (e.g. aldicarb and carbofuran) act as systemic insecticides. A few (e.g. methiocarb) are used as molluscicides for controlling slugs and snails. It is important to distinguish between the insecticidal carbamates and herbicidal carbamates (e.g. propham, chlorpropham) which have only low toxicity to animals.

Carbamate insecticides are formulated in a similar way to OPs, with the most toxic ones (e.g. aldicarb and carbofuran) being available only as granules. They are used, principally, to control insect pests on agricultural and horticultural crops, although they also have some employment for control of nematodes (i.e. as nematicides) and molluscs (i.e. as molluscicides).

1.2.9 *Pyrethroid Insecticides*

Naturally occurring pyrethrin insecticides which are found in the flowering heads of *Chrysanthemum* spp provided the model for the development of synthetic pyrethroids. Synthetic pyrethroids are, in general, more stable chemically and biochemically than are natural pyrethrins. Pyrethroids are solids of very low water solubility which act as neurotoxins in a way similar to DDT (see Chapter 7). They are esters formed between an organic acid (usually chrysanthemic acid) and an organic base (see Figure 1.6). Although pyrethroids are more stable than pyrethrins, they are readily biodegradable, and do not have long biological half lives. They can, however, bind to particles in soils and sediments and show some persistence in these locations. With their low water solubilities, they do not show significant systemic properties, and are not used as systemic insecticides. The hazards that they present relate mainly to short-term toxicity. However, it should be emphasized that they are highly selective between insects on the one hand and mammals and birds on the other. The main environmental concerns relate to their toxicity to fish and non-target invertebrates.

Pyrethroids are formulated mainly as emulsifiable concentrates for spraying. They are used to control a wide range of insect pests of agricultural and horticultural crops throughout the world, and are coming to be used extensively to control insect vectors of disease (e.g. tsetse fly in parts of Africa).

1.2.10 *Phenoxy Herbicides (Plant Growth Regulator Herbicides)*

These constitute the single most important group of herbicides. Familiar examples are 2,4D, MCPA, CMPP, 2,4,DB, and 2,4,5T (for general formula see Figure 1.7). They act by disturbing growth processes in a manner akin to that of the natural plant growth regulator 'indole acetic acid'. They are derivatives of phenoxyalkane carboxylic acids. When formulated as alkali salts they are highly water soluble; when formulated as simple esters they are lipophilic and of low water solubility.

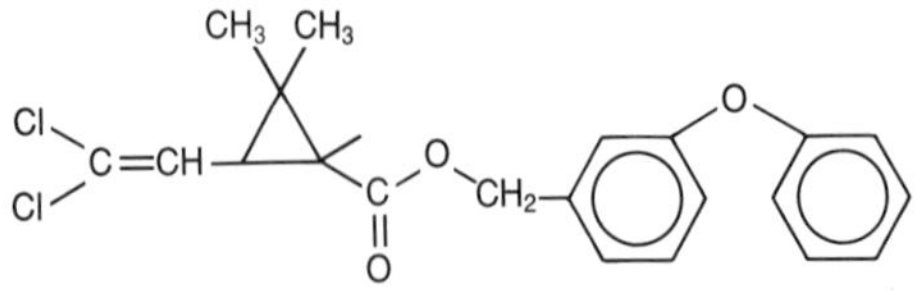

Permethrin

Figure 1.6 Pyrethroid insecticides. These have low polarity and limited water solubility. They are related in structure to the natural pyrethrins, which are also toxic to insects.

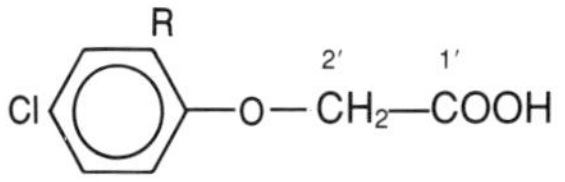

R = CH₃ or Cl

Phenoxyacetic acids (general formula)

Figure 1.7 Phenoxyalkanoic acid herbicides. The example given is a general formula for phenoxyacetic acids, which include 2,4D and MCPA. There are also phenoxypropionic acids (e.g. CMPP) and phenoxybutyric acids (e.g. 2,4DB). All of them have plant growth regulating properties and have some resemblance to the natural growth regulator indole acetic acid.

Most phenoxy herbicides are readily biodegradable and so are not strongly persistent in living organisms or in soil. They are selective, i.e. selectively toxic between monocotyledonous and dicotyledonous plants. Their principal use is to control 'dicot' weeds in 'monocot' crops (e.g. cereals and grass). The environmental hazards are of two kinds. First, there is the problem of unwanted phytotoxicity as a consequence of spray or vapour drift. Second, formulations of certain herbicides of this type have sometimes been contaminated with the highly toxic compound TCDD, e.g. 'Agent Orange', a formulation containing 2,4D and 2,4,5T which was used as a defoliant in Vietnam (see section 1.2.3).

Water soluble salts (e.g. Na^+, K^+) are formulated as aqueous solutions (aqueous concentrates) while lipophilic esters are formulated as emulsifiable concentrates. The latter have sometimes caused unwanted phytotoxicity due to the volatility of some of the esters (vapour drift).

1.2.11 Anticoagulant Rodenticides

For many years, the compound Warfarin (Figure 1.8) has been used as a rodenticide. It is a lipophilic molecule of low water solubility which acts as an antagonist to vitamin K (Chapter 7). More recently, as wild rodents have developed resistance to warfarin, a number of second-generation anticoagulant rodenticides (sometimes called Super Warfarins) have been marketed, which are structurally related to warfarin. These include diphenacoum, bromadiolone, brodiphacoum and flocoumafen and resemble warfarin in their general properties but are more toxic to mammals and birds and are markedly persistent in the livers of vertebrates. Thus, they may be transferred from rodents to the vertebrate predators and scavengers that feed upon them. Owls, for example, have been found to contain residues of them in the UK. Rodenticides are usually incorporated into bait, which is then placed in buildings or out of doors, where it will be taken by wild rodents.

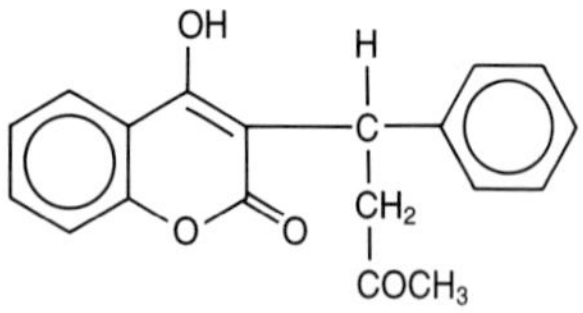

Warfarin

Figure 1.8 Rodenticides. Warfarin and related compounds, such as diphenacoum and brodifacoum, are anticoagulant rodenticides. They are complex molecules bearing some structural resemblance to vitamin K. Their toxic action is due to competition with vitamin K in the liver (vitamin K antagonism).

Anionic

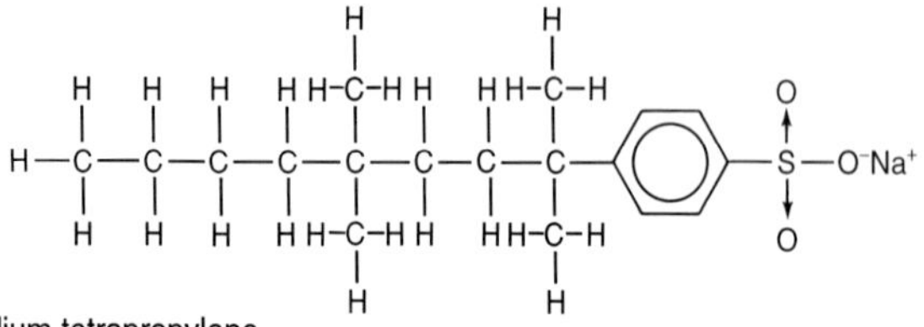

Sodium tetrapropylene benzene sulphonate (hard)

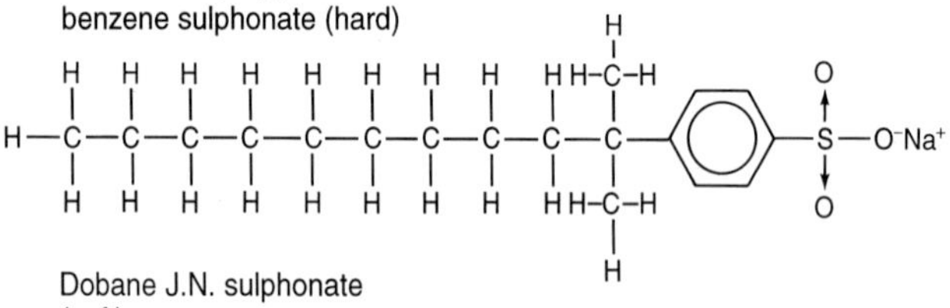

Dobane J.N. sulphonate (soft)

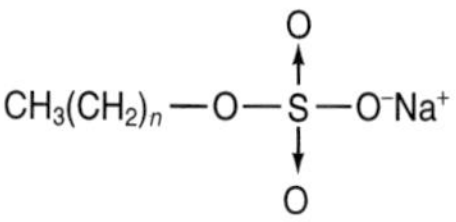

$CH_3(CH_2)_n$—O—S—O⁻Na⁺   Sodium alkyl sulphonate (soft)

Cationic

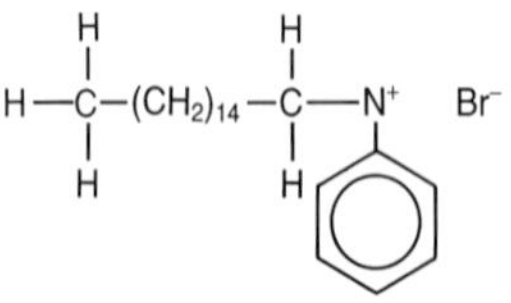

Cetyl pyridinium bromide

Nonionic

Polyglycol ethers of alkylated phenols e.g. lissapol N. stergene $HO-(C_2H_4O)_n$—⬡—R   R = alkyl group

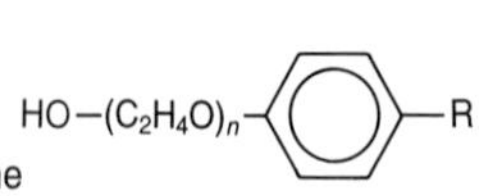

Figure 1.9 Detergents. Detergents are molecules that have both polar and non-polar elements. They may have permanent negative charge (anionic detergents), permanent positive charge (cationic detergents) or a collection of small positive and negative charges over their structure (non-ionic detergents).

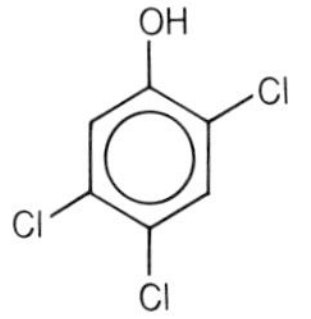

2,4,5-Trichlorophenol

Figure 1.10 Chlorinated phenols. These have acidic properties, releasing H^+ ions when they dissolve in water. They can interact to form dioxin (see section 1.2.3).

1.2.12 *Detergents*

Detergents are organic compounds which have both polar and non-polar characteristics. They tend to exist at phase boundaries, where they are associated with both polar and non-polar media. Some examples are shown in Figure 1.9. Detergents are of three types: (1) anionic, (2) cationic, (3) non-ionic. The first two types have permanent negative or positive charges, attached to non-polar (hydrophobic) C–C chains. Non-ionic detergents have no such permanent charge; rather, they have a number of atoms which are weakly electropositive and electronegative. This is due, in the examples shown, to the electron-attracting power of oxygen atoms.

Detergents are very widely used in both domestic and industrial premises. The major entry point into water is via sewage works into surface waters. They are also used in pesticide formulations, and for dispersing oil spills at sea.

1.2.13 *Chlorophenols*

A number of different polychlorinated phenols (PCPs) occur as environmental pollutants (Figure 1.10). A major source is pulp mill effluent, where they arise because of the chemical action of chlorine (used as a bleaching agent) upon phenolic substances present in wood pulp. Pentachlorophenol is used as a wood preservative, and this is an important source of pollution.

Chlorinated phenols have acidic properties, and are water soluble, chemically reactive and of limited persistence. The tendency of some PCPs to interact and form PCDDs was discussed in section 1.2.3.

1.3 Organometallic Compounds

Some metal ions are so insoluble that they are not toxic if ingested by animals. Liquid metal mercury for example can be swallowed by humans with little effect. Indeed, until the last century, drinking liquid mercury was recommended as a cure for constipation! The low toxicity of tin is demonstrated by its use as a lining in food containers.

Nevertheless, the toxicity of several metals is greatly enhanced if they become bound either deliberately or accidentally to an organic ligand. This changes their chemical behaviour in the environment and within organisms. Metals are modified in this way to increase their toxicity for use as pesticides. Organomercury compounds were used widely as antifungal seed dressings in the UK until as recently as 1993. Organolead compounds have been used extensively for control of caterpillars on fruit crops. Organotin compounds, particularly tributyltin, are extremely toxic. Their main use is for preserving timbers from the activities of aquatic boring animals, and as a component of anti-fouling paints which are applied to the outer surface of boats and fish cages to inhibit settlement by marine organisms. When these substances leach into the environment they can affect non-target organisms. Tributyltin for example has devastated populations of the dog whelk, *Nucella lapillus*, near sites of boating activity in many countries (see section 11.4).

A tragic example of the effects of organomercury compounds occurred in Minimata Bay, Japan, in the 1950s. Metallic mercury released from a paper factory on the shores of the Bay was methylated in the sediments by bacteria to form methyl mercury (see Kudo *et al.*, 1980). Mercury in its methylated form is much more bioavailable than liquid mercury and passed rapidly along the food chain until it reached high concentrations in fish. The local people relied heavily on locally caught fish and were thus vulnerable to poisoning. About 100 people died and many suffered severe disabilities from mercury poisoning. Such incidents are most severe when a local population is highly dependent on a single food source but are rare in more developed regions where food is obtained from much wider sources. Similar problems are likely to occur in the near future in the Amazon Basin. Huge quantities of mercury are being dumped into the river as a by-product of gold refining, and there is evidence that this is becoming methylated and is passing into food chains (Pfeiffer *et al.*, 1989).

1.4   Radioactive Isotopes

1.4.1   Introduction

Since the development of nuclear energy and atomic weapons, there has been an ongoing debate as to the safety of low levels of radioactivity in the environment. We are all exposed to background radiation from cosmic rays and the natural decay of radioactive isotopes. Some consider that this exposure is beneficial as it promotes natural DNA repair mechanisms (a type of 'immunization'). Others consider that there is no safe level of radiation. The contribution of different sources to overall natural background radiation depends to a large extent on local geology. One of the most important sources is radon gas, which may reach levels that give cause for concern in poorly-ventilated houses, especially if they are sited on igneous rocks (Mose *et al.*, 1992).

Three factors determine whether or not radioactive isotopes are harmful to organisms. First, the *nature* and *intensity* of the radioactive decay in terms of the mass and energy of the particles produced. Second, the *half life* of the isotope. Third, the *biochemistry* of the radioactive element. Regarding biochemistry, radioactive isotopes of essential elements will follow the same pathways as their stable forms and accumulate in particular organs. Furthermore, some radioactive non-essential elements may be biochemical analogues of essential elements and follow similar routes in living tissues.

1.4.2 *The Nature and Intensity of the Radioactive Decay Products*

When an atom of a radioactive substance decays it can produce one of four types of particle: alpha (α) beta (β), gamma (γ) or neutrons. It can subsequently decay one or more times until the atom is stable.

The intensity of a radioactive substance is measured in the SI unit of becquerels (Bq) and represents the number of atoms that disintegrate per second. Formerly, radioactivity was measured in curies (Ci) and was equal to the number of disintegrations per second of 1 g of radium. 1 Ci = 3.7×10^{10} Bq and 1 Bq = 2.7×10^{-11} Ci.

The alpha particle consists of two protons and two neutrons and is a positively charged helium nucleus. Alpha particles are relatively massive in comparison to other radioactive emissions. Although they travel only a few centimetres in air, and in biological tissue only a few millimetres at most, their large mass makes them very damaging if they collide with cells, especially if inhaled into the lungs of vertebrates.

The beta particle is an electron and is negatively charged. Beta particles have more penetrating ability than alpha particles with a range of a few metres in air. Sources of beta radiation can be shielded by a thin layer of perspex. However, their small mass means that they do much less damage than alpha particles to tissues.

Gamma rays are quanta of electromagnetic radiation. They are highly penetrating and can pass through several centimetres of lead. As a rule of thumb, the damage they cause is similar to that of beta particles.

Neutrons have no charge and are liberated only when certain elements are bombarded with alpha or gamma rays. They react with other elements only by direct collision. Production of neutrons is the basis for nuclear fission in a reactor. They have a range of several centimetres of lead.

In biological terms, if we were to measure the radioactivity of a substance only in becquerels we would get little information about the effects it might have on tissues. This is because the becquerel takes no account of the nature of the nuclear disintegration, merely its frequency. Two other SI units are needed, the gray and the sievert.

The *gray* is equal to the amount of radiation causing 1 kg of tissue to absorb 1 joule of energy. However, different kinds of radiation do different amounts of damage to living tissue for the same amount of energy. This is sometimes difficult to understand, and a simple analogy is helpful. If a heavy-weight boxer were to gently tap your chin with his fist 100 times and impart X joules of energy each time, you would find this irritating but no long-term damage to your jaw would result. However, if the boxer was to impart 100 times X joules of energy in a single punch, then a broken jaw and several hours of unconsciousness would result. The boxer has imparted the same amount of energy to your chin, but the rate at which it is applied affects the amount of damage that results. Thus, another SI unit, the sievert, is needed. The *sievert* takes account of the different ways in which the same amount of energy can be imparted to tissues. The 'safe' annual exposure of the general public is usually given as 5 mSv. A dose of 20 Sv (a dose received by several hundred workers involved in the Chernobyl clean up – Edwards, 1994) is equal to 20 Gy of beta of gamma emissions or only 1 Gy of alpha particles. Thus alpha particles have about 20 times the effect of beta or gamma radiation for the same number of grays.

1.4.3   Half Lives

Another important consideration is the half life of radioactive isotopes. This is the time taken for half of the atoms of a radioactive isotope to decay. The decay curve follows an exponential decline. For example, ^{32}P is a radioactive isotope of phosphorus used extensively in experiments on molecular biology and plant physiology. It has a half life of 14 days. If we were to start on day 0 with 1 g of ^{32}P, on day 14 we would have 0.5 g of ^{32}P and 0.5 g of ^{32}S (stable sulphur). On day 28 we would have 0.25 g of ^{32}P and 0.75 g of ^{32}S and so on.

After the passage of 10 half lives, the radioactivity of an isotope is no longer significantly different from the background level and is considered to be 'safe'. Thus ^{32}P can be discarded with normal refuse providing that it is held in shielded conditions for 140 days prior to disposal.

The method of disposal of radioactive waste is dictated by the half life of the longest lived isotope. Thus if short-lived isotopes can be separated from long-lived ones before disposal, the volume of waste that has to be held in long-term storage can be greatly reduced. This is much easier in a laboratory or hospital where different isotopes can be separately maintained. Low-level waste is usually placed in concrete and buried in surface trenches. However, medium- and high-level radioactive waste from nuclear reactors contains a complex 'cocktail' of highly radioactive isotopes, some with very long half lives indeed. The only option in this case is secure long-term storage underground. In the UK, the long-term aim is to bury the waste deep underground

Table 1.4 Half lives of some radioactive isotopes

Isotope	Half life
Plutonium-241	13 years
Plutonium-239	24 000 years
Iodine-131	8 days
Iodine-129	160×10^6 years
Strontium-90	28 years
Caesium-137	30 years

although, at the time of writing, the exact location of such a store(s) is still the subject of much political debate. Some examples of the wide range in half lives of different isotopes are given in Table 1.4.

1.4.4 Biochemistry

Some radioactive isotopes are especially dangerous because they follow the same biochemical pathways in the body as stable elements. For example, more than 80 per cent of the total iodine in the human body is contained in the thyroid gland where it forms an essential component of the growth hormone thyroxin. If radioactive iodine is ingested, it becomes concentrated in the thyroid gland and thyroid cancer may result. There is evidence that thyroid cancer has shown a dramatic increase in people living in the vicinity of Chernobyl (Kazakov *et al.*, 1992). The radioactive isotope strontium-90 (Table 1.4) follows the same pathways as calcium in humans. Most people in the world contain trace amounts of this isotope in their bones following its global dispersion after atmospheric atom bomb tests in the 1950s and 1960s. Caesium-137 follows potassium pathways and has been a particular problem in areas subjected to Chernobyl fall-out, such as Northwest England (Crout *et al.*, 1991; Simkiss, 1993) and Scandinavia (Åhman and Åhman, 1994). Where nutrients are strongly recycled by the vegetation, it may take many decades before levels of Chernobyl contamination decline to background levels.

1.5 Gaseous Pollutants

The most important gaseous pollutants are ozone (O_3), and oxides of carbon, nitrogen and sulphur.

On a global level, the main concern with ozone is the reduction in its concentration in the upper atmosphere. The well-publicized ozone 'hole' which occurs over the Antarctic (but now detected at high latitudes in the Northern Hemisphere), is caused by the degradative effects of chlorofluorocarbons

(CFCs) on ozone molecules (see Chapter 3). CFCs are released from aerosol containers, from the coolants in domestic refrigerators when they are broken up, and from foam packaging. The ozone layer absorbs ultraviolet light, so that one hazard associated with its thinning is an increase in the rates of skin cancer. Environmentally it has been suggested that the increased radiation could decrease photosynthesis of phytoplankton in the Antarctic.

At a local level, ozone is produced in photochemical smogs where oxides of nitrogen from car exhausts (NO, NO_2, sometimes known as 'NO_x' or 'NOX' gases) and fumes from other fossil fuel consumption react with moisture under the action of sunlight (Bower *et al.*, 1994). High concentrations of ozone in the air irritate respiratory epithelia of animals and can directly affect the growth of some plants (Mehlhorn *et al.*, 1991). Tobacco is particularly sensitive to ozone-induced damage (Heggestad, 1991). Ozone has been considered a major factor in forest dieback in Germany (Postel, 1984).

Levels of carbon dioxide rarely reach toxic concentrations except in very confined places. The effects of CO_2 on global warming will be discussed in Chapter 15.

Sulphur dioxide is produced mainly from volanoes and fossil fuel burning. The SO_2 dissolves in water droplets, forming sulphurous and sulphuric acids which fall as 'acid rain'. Rain with a low pH may directly damage the leaves and roots of plants. Furthermore, nutrients may be washed out of acidified soils as the hydrogen ions displace essential elements from soil particles. Plants growing on such soils may become deficient in one or more trace elements. A deficiency of magnesium, for example, is thought to be one of the possible causes of forest dieback in Germany. Acid rain may also increase the mobility of metal pollutants in soils if the pH drops sufficiently low (see Figure 4.3). The effects of acid rain are much greater on soils and lakes with a low buffering capacity. One of the reasons why Scandinavia has been so badly affected by acid rain is the poor buffering capacity of the soils developed from the granite bedrock which underlies much of the region. The effects of acid rain on ecosystems are discussed in section 14.3.

Further Reading

MANAHAN, S. E. (1994) *Environmental Chemistry*, 6th Edition A detailed and comprehensive text covering both inorganic and organic chemicals. Two useful chapters on basic principles for those with limited background in chemistry.

BUNCE, N. (1991) *Environmental Chemistry* Gives an account of the pollutants discussed here; very little, however, on pesticides and radionuclides.

MERIAN, E. (ed.) (1991) *Metals and their Compounds in the Environment* Almost everything you could possibly want to known about metals in this comprehensive text (1438 pages!).

PAASIVIRTA, J. (1991) *Chemical Ecotoxicology* Gives a brief overview of most of the important organic and inorganic pollutants.

GUTHRIE, F. E. and PERRY, J. J. (1980) *Introduction to Environmental Toxicology* A multiauthor work giving an in-depth account of most of the important organic and inorganic pollutants.

HASSALL, K. A. (1990) *The Biochemistry and Uses of Pesticides* Gives a very readable account of the chemical properties of the major types of pesticide.

ÅHMAN, B. and ÅHMAN, G. (1994) A very interesting study of the consequences of Chernobyl fall-out in reindeer, and the local human population which rely on them.

EDWARDS, T. (1994) An excellent article which captures the impact of Chernobyl in the usual National Geographic style.

JUKES, T. (1985) An interesting discussion of the consequences of the narrow window of essentiality for selenium.

NIEBOER, E. and RICHARDSON, D. H. S. (1980) A classic influential paper on the chemistry of metal ions which had a major impact on studies of metal toxicity.

2

Routes by Which Pollutants Enter Ecosystems

Pollutants may enter ecosystems through:

'unintended' release in the course of human activities (e.g. in mining operations, shipwrecks and fires);
disposal of wastes (e.g. sewage, industrial effluents);
deliberate application of biocides (e.g. pest control and vector control).

Some of the chemicals released in this way can also reach unusually high levels locally due to natural processes such as weathering of rocks (many metals and inorganic anions) and volcanic activity (SO_2, CO_2, and aromatic hydrocarbons). As noted in the Introduction, there are problems of defining what actually constitutes pollution. Some authorities prefer to restrict the terms 'pollution' and 'pollutants' to the consequences of human activity. However, it is often difficult or impossible to determine the relative contribution of human processes and natural ones to residues that are present in the general environment.

2.1 Entry into Surface Waters

The discharge of sewage into surface waters represents a major source of pollutants globally (Table 2.1). Domestic wastes are discharged mainly into sewage systems. Industrial wastes are discharged *either* into the sewage system *or* directly into surface waters.

The quality of the sewage that is discharged into surface waters depends on (1) the quality of the raw sewage received by the sewage works, and (2) the treatment of the sewage that takes place within the works (Benn and McAuliffe, 1975). Urine, faeces, paper, soap and synthetic detergents are important constituents of domestic waste. Industrial wastes are many and varied and their quality depends on the nature of the operations that are followed in particular establishments. A variety of treatments may be carried out at

Table 2.1 Major routes of entry to surface waters

Route	Major pollutants	Comments
Sewage outfalls	A very wide range of organic and inorganic pollutants from commercial and domestic sources; detergents generally present	Highly variable; dependent not only on what sewage works receive but also on the treatments that sewage is given
Outfalls from commercial premises	Dependent on the commercial activity; wide range of pollutants from chemical industry; heavy metals from mining operations; pulp mills an important source of pollutants in some areas	Concentration of pollutants in effluents must stay below statutory limits
Outfalls of nuclear power stations	Radionuclides	Subject to regular monitoring and close control in most countries
Run-off from land	A variety of pollutants dumped on land surface; pesticides	Generally uncontrolled and difficult to measure

From the air	(a) Precipitation with rain or snow	Sometimes pollutants transported over great distances
	(b) Direct application of biocides	Control of pests, parasites, vectors of disease and aquatic weeds
	(c) Accidental contamination by sprays or dusts	Aerial spraying a potential problem
Dumping at sea	Raw sewage; radiochemicals and toxic wastes in sealed containers dumped in deep ocean	Concern sometimes expressed about release from containers in the longer term when they degenerate
Release from oil rigs and terminals	Hydrocarbons	Sometimes accidental, sometimes as a result of war (e.g. Gulf War in Kuwait)
Shipwrecks	Hydrocarbons and some other organic pollutants	Wrecks of oil tankers particularly a problem

sewage works to improve the quality of sewage before it is discharged into surface waters (Figure 2.1). In primary treatment, sewage is passed through a sedimentation tank where it has a retention period of several hours. At this stage, 'primary sludge' will settle out. Subsequent to this, during secondary treatment, biological oxidation and flocculation of most of the remaining organic matter takes place. Typically this is done by either the 'activated sludge' process or by 'biological filtration'. Conversion of ammonia to nitrites and nitrates by microorganisms is a feature of the process. Also, detergents are removed by biological oxidation. Much of the organic matter entering sewage works is converted into sludge – and this is disposed of by spreading on land as a fertilizer or by dumping on land or at sea. The sewage effluent resulting from secondary treatments may be subjected to further treatments to remove constituents such as phosphate, nitrate, silicates and borates, depending on the quality of the final effluent that is required.

Important properties of sewage are levels of suspended solids, chemical oxygen demand (COD) and biochemical oxygen demand (BOD). COD is a measurement of the amount of oxygen required to achieve a complete chemical oxidation of one litre of a sewage sample. BOD is a measurement of the amount of dissolved oxygen used by microorganisms to oxidize the organic matter in one litre of a sewage sample. Sewage that is discharged into surface waters should have values for COD and BOD that fall below agreed limits. If the values are too high, this means that the organic content of the sewage is too high for discharge into receiving waters. Such a discharge could cause

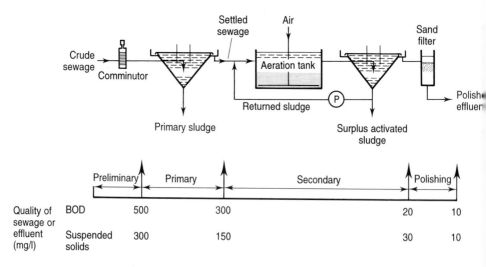

Figure 2.1 Conventional treatment of sewage by the activated sludge process. The top of the diagram illustrates typical stages in sewage treatment. The lower figure indicates the quality of sewage at different stages of treatment. After Benn and McAuliffe (1975), p. 80, with permission.

substantial reduction in the oxygen level of the water with serious consequences for aquatic organisms. In practice, the quality of sewage effluent varies enormously from country to country and from place to place. Because of the high cost of sewage treatment, no more is done than the situation requires, even in developed countries. Full advantage is taken of the capacity of receiving waters to achieve degradation of sewage components. Sewage is a rich source of organic and inorganic pollutants, prominent among them detergents which are extensively used in both domestic and industrial premises. These detergents have given rise to serious pollution problems. For further discussion of this question see Benn and McAuliffe (1975).

The type of pollutants in industrial effluents depends largely on the industrial processes that are being followed. Heavy metals are associated with mining and smelting operations, chlorophenols and fungicides with pulp mills, insecticides with moth-proofing factories, a variety of organic chemicals with chemical industry and radionuclides with atomic power stations. In developed countries there are close controls over the permitted levels of release of chemicals in industrial effluents. Offshore industrial activities, such as oil extraction and manganese nodule extraction, lead to the direct discharge of pollutants to the sea.

Apart from direct discharge, pollutants are sometimes dumped into surface waters at considerable distances from the premises where they are produced. Such dumping is largely restricted to the sea. Sometimes sludge from sewage works is taken well out to sea and dumped there. Also, radioactive wastes and chemical weapons have been dumped at sea in sealed containers. Questions have been asked about release in the longer term, because these containers will eventually disintegrate. It is usual to dump dangerous wastes where the sea is deep, to minimize risk of contamination of surface layers of the ocean.

A further problem is the release of oil from tankers, most dramatically in the case of shipwrecks, when large quantities are discharged in a relatively short time in one area. However, it is important to put such incidents as these – and the disastrous release of oil during the Gulf War – into a wider perspective. The total input of petroleum hydrocarbons to the marine environment has been estimated at 3.2 million tons per year. Although oil tanker disasters can cause great damage, the input from them ranks below normal tanker operations and discharges from industrial and municipal waste.

Biocides are sometimes deliberately applied to surface waters in order to control invertebrates or plants. Herbicides have been used to control aquatic weeds in lakes and water courses. Insecticides are applied to control fish parasites at fish farms in both freshwater and marine locations and to control pests in watercress beds. Tributyl tin fungicides have been incorporated into antifouling paints for use on boats, and this has led to marine pollution.

Most of the examples given so far have been of deliberate actions which are, at least in theory, carefully controlled. There are, however, many cases of 'accidental' pollution which are not under direct human control. Pollutants present in air may enter surface waters as a consequence of precipitation of dust or droplets or with rain or snow – or simply due to partition from air into water. Pollutants present on the land surface, e.g. metals or pesticides, may be washed into rivers, streams or oceans when there is heavy rainfall. They may be in the free or particulate state, or attached to soil or mineral particles. There is a particular problem with aerial spraying of pesticides, where there is a risk of spray drift into surface waters. Some pesticides are extremely toxic to aquatic organisms, and it is very important that spray droplets should not directly contaminate waters.

Release of pollutants into moving surface waters is followed by dilution and degradation. Consequently, biological effects are most likely to be seen at or near the point of release. Where pollutants enter rivers, there may be a biological gradient downstream from the outfall. Sensitive organisms may be absent near the outfall, but re-appear downstream. In fast-flowing rivers, the dilution effect is marked and pollutants are unlikely to reach high concentrations at a reasonable distance below the outfall.

By virtue of their size, and the action of currents, oceans can effectively dilute incoming pollutants. Of greater concern are lakes and small inland seas. Here, pollutants are brought in by rivers, and other routes. Because they have no effective outlets, pollutants will tend to build up in them as water evaporates, sometimes with detrimental consequences (section 14.3). Much depends upon the rates of degradation or precipitation that will remove pollutants from the water. The pollution of the Great Lakes of North America provides a good example and will be discussed in Chapter 15.

2.2 Contamination of Land

As with pollution of surface waters, contamination of land may or may not be deliberate. Deliberate contamination may involve either the disposal of wastes or the control of animals, plants or microorganisms with biocides. Accidental contamination may be the result of short-term or long-term aerial transport, flooding by rivers or seas, or collision of tankers/lorries carrying toxic chemicals (Table 2.2).

The dumping of wastes at landfill sites is a widespread practice. Indeed, many old sites could be regarded as ecological 'time bombs'. On the one hand there is the question of disposal of domestic and general industrial wastes. On the other hand there are toxic wastes which require more careful handling. Of particular concern are radioactive wastes from nuclear power stations. With the latter there are stringent regulations for safe disposal, to minimize contamination of the land surface, and neighbouring surface waters. One practice is to embed the disposed radioactive material in concrete.

The use of sewage sludge as fertilizer on agricultural land is another source of pollution. Heavy metals, nitrates, phosphates and detergents are all added to soil in this way. Land is also contaminated by aerially transported materials. Smoke and dust from chimneys can fall on neighbouring land,

Table 2.2 Major routes of contamination of land

Route	Major pollutants	Comments
Waste dumping including rubbish dumps/landfall sites/industrial dumps	A very wide range of different pollutants	Some industrial dumps are high in particular pollutants e.g. oil, metal ore deposits, PCBs etc.
Application of pesticides to agricultural land and forests	Insecticides, rodenticides, herbicides and fungicides as sprays, dusts, seed dressings etc.	In most countries there are strict regulations controlling the application of pesticides
Control of insect vectors of disease	Insecticides	Major pollution over large areas as a consequence of control measures against malarial mosquito and tsetse fly
Application of sewage to agricultural land	Heavy metals, nitrates, detergents	
Flooding by rivers or seas	A variety of pollutants including those associated with sewage	
Precipitation from air as dust or droplets or in rain or snow	Pollutants associated with soot and dust, acid rain, pesticides	Transport may be over short distances (spray drift, soot and dust from chimneys) or long distance (brought down especially by rain and snow)

carrying with them a variety of organic and inorganic pollutants. Gases such as sulphur dioxide, nitrogen oxides and hydrogen fluoride released from chimneys cause damage to vegetation in the neighbourhood of industrial premises. Thus, pollution of the land surface may occur in the immediate vicinity of domestic and industrial premises which cause air pollution. Additionally, as with surface waters, pollutants reach the land after travelling considerable distances. They are carried down by rain or snow, in solution or suspension or with associated dust particles.

Land is sometimes also contaminated by pollutants when there is flooding by rivers or seas. Considerable areas of the land surface are treated with biocides, in order to control vertebrate and invertebrate pests, plant diseases, weeds and vectors of disease. This is important in areas of intensive agriculture, where a variety of different pesticides are used during the course of a farming year. Pesticides are applied as different formulations – as sprays, granules, dusts and seed dressings. The manner of application and the nature of the formulation influence the way the pesticides are distributed in the crop and in the soil. There is a potential problem of spray drift to areas outside the target sites, during the course of pesticide application. This is particularly so with aerial spraying, and is very dependent upon the strength and direction of winds at the time of the operation. There is sometimes also a problem of pesticides moving through soil to contaminate groundwater, especially where there are cracks in the soil profile, allowing rapid percolation of water. A field experiment conducted by MAFF at Rosemaund, England, demonstrated that a variety of pesticides can find their way into drains and water courses following heavy rain.

Apart from agricultural applications, pesticides are applied over large areas for other purposes. Thus, insecticides are used extensively in Africa to control tsetse fly and locust swarms. Aerial spraying of insecticides has taken place over forests in Canada to control insect pests (see Chapter 15), and over nesting colonies of *Quelea* (a bird pest) in parts of Africa.

2.3   Discharge into the Atmosphere

Pollutants enter the atmosphere in the gaseous state, as droplets or particles, or in association with same. When in the gaseous state they may be transported over considerable distances, with the movements of air masses. Particles and droplets, on the other hand, are more likely to move over only relatively short distances before falling to the ground. However, they too may sometimes undergo long-distance transport especially when they are of small diameter.

Chimneys of both industrial and domestic premises are important sources of atmospheric pollution (Table 2.3). Carbon dioxide (CO_2), sulphur dioxide (SO_2), oxides of nitrogen (NO_x), hydrogen fluoride and chlorofluorocarbons

Table 2.3 Major points of entry into the atmosphere

Route	Major pollutants	Comments
Domestic chimneys	Many organic compounds, including hydrocarbons associated with smoke particles or as vapours. SO_2, CO_2, NO_2 and other gases	Level of pollution dependent upon quality of fuel burned and clean-up of flue gases
Chimneys of industrial premises, power stations, etc.	As for domestic chimneys but also many other pollutants depending on the practices followed on the site[a] Radiochemicals from nuclear power stations	With hazardous substances, procedures for cleaning up effluent gases are very important
Internal combustion engines and jet engines	CO_2, (NO_x) hydrocarbons and other organic pollutants; lead compounds (mainly inorganic but some organic lead)	Level of pollution very dependent upon design of engine and exhaust system; the growing use of lead-free petrol is restricting lead pollution
Pesticide applications	Insecticides, fungicides and herbicides	Volatile pesticides enter the air in the vapour state also droplets of pesticide spray and pesticide dust formulations reach the atmosphere
Escape from refrigerators	Chlorofluoromethanes	
Aerosols	Chlorofluoromethanes, e.g. CF_2Cl_2 $CFCl_3$	Many countries now have strict controls over the use of these compounds as propellents

[a] *Note*: For further details, see Chapter 2.

35

(CFCs) are examples of gases released in this way. The combustion of fuels releases CO_2, SO_2, NO_x and a variety of organic compounds (e.g. PAHs) which are products of incomplete combustion. The level of pollution depends on the quality of the fuel and the extent to which flue gases are cleaned up. Some forms of coal (e.g. brown coal from Silesia) are high in sulphur and can cause very serious pollution with SO_2. Many of the organic compounds released from chimneys are present in smoke particles. The subsequent movement of pollutants is dependent upon atmospheric conditions and on the height and the form of the chimney releasing them. Under clear and warm conditions pollutants will be quickly diluted due to the mixing of air. As the earth's surface is warmed by sunlight, hot air will rise from the vicinity of the chimney, producing convection currents and carrying pollutants with it. Cold, clean air will flow in to replace it. If there is a side wind, air-borne pollutants will be carried away from the initial point of release, with further dilution. In the evening this process may be reversed, as the air cools. Then, if there is no wind, a layer of mist or fog may form, trapping a well of cold air beneath it (Figure 2.2). The following morning the sun will be unable to penetrate the layer of mist/fog, thus preventing a warming of the air and consequent dispersal of air pollutants. The air pollutants will become trapped in the vicinity of the chimney from which they were emitted. Thus, in general, the dispersal of air pollutants is favoured by warm, dry conditions, with a steady side wind. The dispersal of pollutant will be more effective from high chimneys than from low chimneys. In general, the higher the point of release, the greater the height that pollutants will reach in the atmosphere, and the greater the distance they are likely to travel (Chapter 3).

The internal-combustion engine represents another important source of air pollution. Apart from vehicles on roads, the engines of aeroplanes and ships cause pollution of air and sea. Jet engines are a significant source of air pollution. During the operation of internal-combustion engines, chemical reactions take place which generate substances not originally present in the petrol and air mixture delivered by the carburettor. Carbon monoxide and nitrogen oxides are released, together with a variety of organic molecules which are the products of incomplete combustion. The latter include PAHs, aldehydes and ketones, in addition to the normal constituents of petrol. Petrol is also a source of organolead and inorganic lead compounds which arise from lead tetra-alkyl, used in some petrols as an 'antiknock' (to control semi-explosive burning during the operation of the engine). Nowadays there is a strong movement towards the greater use of lead-free petrols, to reduce the emissions of lead in its various forms from car exhausts.

The control of emissions from internal-combustion engines is a complex subject that will only be considered, in outline, here. In an 'uncontrolled' vehicle, effectively all of the carbon monoxide, nitrogen oxides and inorganic lead compounds, and about 65% of hydrocarbons are released from the exhaust system. Evaporation from the fuel tank and carburettor accounts for a significant loss of hydrocarbons, and of volatile lead tetra-alkyl. Finally, there

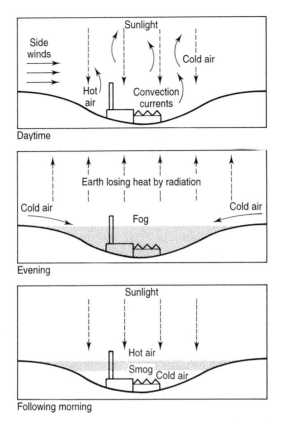

Figure 2.2 Inversion effects in air pollution. After Ben and McAuliffe (1975), p. 80, with permission.

can be a substantial loss of hydrocarbons due to leakage around the pistons and into the crankcase (crank case 'blowby').

With modern engines there have been considerable improvements in design which reduce all these sources of pollution. Exhaust emissions are substantially reduced by the incorporation of catalytic converters and filters into the exhaust system. These remove nitrogen oxides, carbon monoxide and hydrocarbons. A problem with catalytic converters is that they are rapidly poisoned by tetra-alkyl lead. Recognition of this has hastened the phasing out of leaded petrols. Further improvements in exhaust emission have come through the optimization of engine performance (critical factors are air-fuel ratios, ignition timing and cylinder design). Crank case 'blowby' has been reduced by recycling 'blowby' gases via the carburetter system. Finally, evaporative losses from carburettor and fuel tank have been reduced by improvements in design.

Speaking generally, considerable improvements in the control of the release of pollutants by internal-combustion engines have been effected in recent years. However, with the rising use of motor vehicles, this remains one of the major sources of air pollution. The extent to which improvements have been made varies greatly from country to country. The standards that currently operate in Western Europe, North America and Australia are not usually found in other parts of the world. Recently, there has been considerable concern about the release of particulates from diesel engines (Tolba, 1992).

Air pollution also arises because of the use of pesticides. The application of pesticides as sprays or as dusts is not a very efficient process. A substantial proportion of the pesticide that is applied does not reach the crop or the soil surface. Aerosol droplets, dust particles with adhering pesticides, and pesticides in the gaseous state pass into the air. This is a particularly difficult problem when pesticides are applied aerially. Climatic factors influence the extent to which pesticides contaminate the atmosphere. Strong side winds tend to move them away from original areas of application, with the risk that neighbouring areas lying downwind will be contaminated. Volatilization is most rapid where air temperatures are highest. Thus, pesticides show a greater tendency to volatilize into the air under tropical conditions than they do under temperate conditions. This point needs to be borne in mind when attempting to extrapolate from field studies performed in the temperate zone to make predictions about pesticide fate under tropical conditions.

Another factor of importance is droplet size. Very small spray droplets produced during low volume spraying fall more slowly to the ground than large droplets, because their sedimentation velocity is slower, and are liable to travel for relatively long distances before reaching the ground. In general, environmental factors such as wind speed, temperature and humidity need to be taken account of when planning spray operations in order to maximize the amount of pesticide reaching its target, and to minimize air pollution.

Radiochemical pollution of the air due to the explosion of atomic devices on or above the land surface was a problem for many years after the last war. By international agreement however, this practice has now been discontinued, but concern remains over accidental release from establishments such as power stations and atomic research stations which handle nuclear materials. The seriousness of the problem was clearly illustrated by the Chernobyl accident of 1986 in the Ukraine, when a nuclear reactor caught fire with consequent widespread air pollution with radionuclides (see Chapter 1). Half of the reactor contents were dispersed.

An important group of air pollutants are low-molecular-weight halogenated hydrocarbons, such as the chlorofluorohydrocarbons (CFCs), which are used as propellants and in refrigerators, and chlorinated compounds (e.g. CH_2Cl_2) which are used for dry cleaning. These volatile substances can escape into the air during normal usage and after waste disposal. A major problem with CFCs is that they can reach zones of the upper atmosphere where they can cause damage to the ozone layer (ozonosphere) (Chapter 3).

Many of the pollutants found in air exist in the same forms that were originally released from the land, or water surface. There are, however, some pollutants that are generated by chemical reactions within the atmosphere. This happens, for example, in the case of the photochemical 'smog' which has caused problems in Los Angeles, and other cities where there are large numbers of vehicles, and high solar radiation. Under these conditions, if there is no wind, nitrogen dioxide and organic compounds released from car exhausts together with oxygen are involved in a complex series of reactions. The products include ozone, and organic compounds such as peroxyacetyl nitrate (an eye irritant).

2.4 Quantification of Release of Pollutants

Legislation which has the purpose of controlling pollution focuses on the *amounts* of pollutants that may enter the environment, and the *rates* at which they may be released. For major pollutants international agreements are necessary to control the input. In the case of carbon dioxide the UN conference on Environment and Development at Rio de Janeiro in June 1992 recommended reducing carbon dioxide emissions to 1990 levels by the year 2000. Even this modest goal was not supported by the USA, the largest producer of CO_2. A more recent conference in Berlin (April 1995) recommended specific targets for a longer term (up to the year 2020) with a two-year timetable to negotiate the final figures. Nevertheless, with the cautious approach of the USA and opposition of the oil-producing countries the outcome is uncertain. More has been achieved in agreements about CFCs which industrialized countries agreed to phase out by the year 2000 (a 10-year time lag was allowed to developing countries). Subsequently this date advanced to 1996. In Europe the EEC has been active in negotiating reductions of SO_2 and NO_x to limit the effects of 'acid rain' (Table 2.4).

In ecotoxicity testing of new industrial chemicals the testing protocols are influenced by the amount of chemical produced per annum since the amount

Table 2.4 Amounts of gaseous pollutants released per year, globally, in tons[a]

	Anthropogenic sources	Natural sources[b]
CO_2	6 000 000 000	100 000 000 000
SO_x	100 000 000	50 000 000
NO_x	68 000 000	20 000 000
CFCs	1 100 000	0

[a] Data from Tolba (1992) and UNEP (1993).
[b] There is considerable uncertainty in the natural sources data.

39

produced per annum gives some indication of the possible scale of any consequent pollution problem. In the current regulations of the European Commission, if production is at a low level – only a minimal base set of tests is usually required. However, when production exceeds certain thresholds, additional testing is required (for further details see Walker *et al.*, 1991b).

Knowledge of the rate and pattern of release of pollutants is necessary when modelling the environmental fate of pollutants (see Chapter 3).

Further Reading

BENN, F. R. and McAULIFFE, C. A. (eds) (1975) *Chemistry and Pollution*, Useful chapters on sewage treatment and major sources of air pollution.

BUTLER, J. D. (1979) *Air Pollution Chemistry*, An authoritative work on air pollution.

SALOMONS, W., BAYNE, B. L., DUURSMA, E. K. and FÜRSTNER, V. (eds) (1988) *Pollution of the North Sea: an Assessment*, Specialized chapters on sources of pollution in the North Sea.

CLARK, R. B. (1992) *Marine Pollution*, 3rd Edition. A standard text giving a straightforward account of marine pollution.

MANAHAN, S. E. (1994) *Environmental Chemistry*, 6th Edition. A great deal of information on the release of pollutants in this comprehensive test.

3

Long-range Movements of Pollutants in the Environment

Pollutants are capable of movement over considerable distances. They can be carried over boundaries between different countries, thereby raising political as well as environmental problems, for the export of environmental chemicals tends not to be appreciated by countries that receive them. Scandinavian countries, for example, have objected to the deposition of aerially transported sulphur dioxide (SO_2) originating from the British Isles. For the most part, transport over large distances is the consequence of mass movements of air or water. However, movement may also be by diffusion, which can be rapid in air, but less rapid in water. Movement by diffusion may be very localized, or may occur over large distances, especially in air.

The general question of movement in water and in air will be discussed before considering models which describe or predict the movement of pollutants.

3.1 Transport in Water

The pollutants present in surface waters exist in diverse states. They may be in solution, and/or in suspension. Suspended material may be in the form of droplets (e.g. oil) or particles and pollutants may be dissolved in droplets or absorbed by solid particles. All of these forms can be transported by water over considerable distances. Particulate material is liable to fall to the bottom of surface waters, e.g. where rivers enter the sea their rate of flow is checked and the coarser transported particles fall to the bottom with estuarine deposits. Liquid droplets may rise to the surface, or be carried down by particles to the sediment, depending on their density. With oil pollution, both of these things occur – light oil rises to the surface, but 'heavy' oil residues go into sediment.

In rivers, pollutants are transported over varying distances. The distances travelled depend on factors such as the stability and physical state of the pollutants and the speed of flow of the river. The distance travelled is likely to be greatest where stable compounds are in solution and where rivers are fast-flowing. In general, the concentration of a pollutant continually falls with increasing distance below an outfall and this may be reflected in the changing composition of the fauna and flora (Figure 3.1). The importance of long-distance transport of pollutants by rivers was clearly demonstrated when the river Rhine became polluted with the insecticide endosulfan in 1969. The initial release was evidently in the middle section of the river, near Frankfurt, but the transported compound was detected by Dutch scientists working near the Rhine estuary some 500 km downstream.

Once pollutants reach lakes or oceans they may be transported by currents. The major oceans of the world are traversed by surface currents, so it is possible for pollutants to be moved from one continent to another. These currents are wind-driven, and they move roughly at right angles away from the direction of prevailing winds. In both the Atlantic and the Pacific oceans there are large circular patterns of currents (gyres) over most of the surface area. The movement is clockwise in the Northern Hemisphere, but counterclockwise in the Southern Hemisphere. The Gulf Stream of the North Atlantic is part of a clockwise gyre system, and brings warm water to the shores of the British Isles.

The density of sea water is an important factor. Water may increase in density due to a fall in temperature or due to an increase in salt concentration (e.g. due to evaporation). When water masses increase in density they move towards the bottom of the ocean. Downward movements are countered by upwards movements of advecting water from the lower levels of the ocean. Deep water circulation in the oceans of the world is illustrated in Figure 3.2.

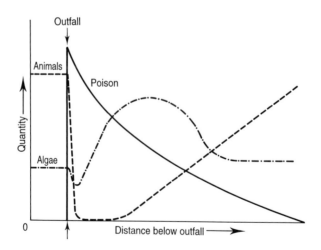

Figure 3.1 Diagram of the effects of poisonous effluent on a river. Quantity (vertical axis) refers to the concentration of chemical, or number of individuals per unit volume in water. After Hynes (1960).

Figure 3.2 Deep water circulation in the oceans. The thick lines are major bottom currents; the thin lines represent the 'return' more general flow of water. From Turekian (1976).

It is sometimes assumed that oceans are so large that dilution will quickly reduce pollutant concentrations to such low levels that they no longer constitute a problem. One shortcoming of this argument is that the distribution of pollutants in oceans is far from uniform. The movement of particulate matter with currents, and its subsequent precipitation, ensures unequal distribution. This problem is readily seen with the precipitation of sediment in estuaries as mentioned earlier. Inshore waters tend to have substantially higher levels of pollution than the open sea.

When persistent pollutants enter marine food chains they can be moved over large distances by migrating animals and birds. Some fish, whales and fish-eating sea birds migrate over many thousands of miles, taking pollutants with them. This can lead to transfer from one ecosystem to another, e.g. where contaminated fish or birds migrate over large distances and are then eaten by vertebrate predators.

3.2 Transport in Air

Traces of persistent pollutants such as organochlorine insecticides and PCBs have been detected in snow and in animals living in polar regions, at locations far removed from any point of release. This clearly illustrates that pollutants can be transported over very large distances by the movement of air masses. The main sink of CFCs is in the stratosphere, indicating the importance of vertical movement through the troposphere for small, stable and volatile molecules. The translocation of pollutants over large distances is dependent upon the physical state of the pollutants, and the movement of air masses.

Consider, first of all, the physical state of the pollutants. Some air pollutants are in the gaseous state. Examples include CO, NO_x, SO_2, HF and small volatile halogenated molecules such as CFCs, trichloroethylene ($CCl_2 = CHCl$) and carbon tetrachloride (CCl_4). These may move through the air by two processes – mass transport and diffusion. The question of mass transport will be returned to shortly. Diffusion of gases may be of two types. First, there is diffusion along a concentration gradient, which proceeds at rates determined by Ficks law of diffusion which states:

$$\theta = -D(C_2 - C_1)$$

where θ is rate of diffusion, D is the diffusion constant and $(C_2 - C_1)$ is the concentration gradient. Thus, net diffusion occurs in a direction that will tend to remove the gradient. The steeper the concentration gradient the faster the rate of movement. Second, there is thermal diffusion. In situations where a thermal gradient exists, 'hot' molecules of high velocity move faster than 'cold' molecules of low velocity.

Apart from the gaseous state, pollutants exist as droplets or particles *or* in association with droplets and particles. The latter situation is commonplace. Particles of dust or soot, or droplets of water can be of complex composition, containing a range of polluting substances. Apart from releases from the earth's surface, pollutants may be incorporated into rain droplets. This may happen during the course of precipitation ('wash-out') or during the formation of droplets in clouds ('rain-out'). Soluble gases such as SO_2 and NO_x tend to dissolve in rain droplets. Also, rain may bring down dust particles present in the air.

Air movements on the global scale are relatively complex. First, there is a layer of air close to the earth's surface (say 1–4 km in height) which is subject to particular turbulence, and localized air flow. Pollutants released within this zone are likely to return to ground quickly, travelling only relatively short distances. On the other hand, pollutants which reach greater heights may be transported over considerable distances, carried by circulating air masses.

The part of the atmosphere relevant to this discussion extends some 35 km above the earth's surface. This is the lower atmosphere, which accounts for about 99% of the total air mass. It is divided into the troposphere (first 10–11 km) and the stratosphere, which lies above it. The troposphere is characterized by strong vertical mixing – individual molecules can move through the entire height in a matter of days. There is little vertical mixing in the stratosphere above it. The boundary between these two layers is termed the tropopause. Within the stratosphere there lies a band of relatively high ozone concentration, termed the ozone layer (ozonosphere).

Within the troposphere there are regular patterns of air circulation, characteristic of different climatic zones (Figure 3.3). Both northern and southern hemispheres are divided into three circulation zones. First, at the equator, there are sharply rising currents of hot air to either side, drawing in flows of cooler air from the north and the south respectively. In the upper part of the troposphere the rising air cools and then moves either northward or southward. This 'poleward' air flow begins to subside and to flow towards the earth's surface at between 20 and 35 degrees latitude. Some of this air will then move in a surface flow towards the equator, thus completing the cycle. The surface winds resulting from this circulation are termed the North-east and South-east trade winds in the northern and southern hemispheres respectively. These terms illustrate the point that the air flow is not directly on a North-to-South axis – there is also a West-to-East component, which is a consequence of the influence of the earth's rotation upon air movement in the troposphere. Between approximately 30 and 60 degrees latitude, there is a reversal of this

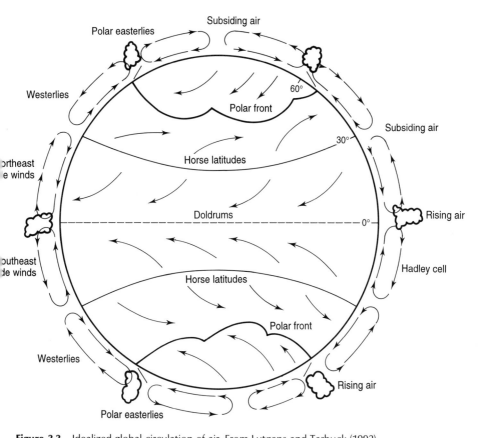

Figure 3.3 Idealized global circulation of air. From Lutgens and Tarbuck (1992).

circulation pattern in both hemispheres. The surface flow is 'poleward' and not towards the equator. Finally, in the polar regions, beyond 60 degrees latitude, air flow is again reversed, as shown in Figure 3.3.

It is clear from this description that air pollutants – including those associated with small particles and droplets – are liable to be transported over large distances once they enter the main air circulation a few kilometres above the earth's surface. It follows that the release of pollutants some distance above the earth – e.g. from aeroplanes, or from high chimneys – may lead to long-distance transport. Pollution problems may then become 'global' rather than local. Pollutants may be brought down by rain or snow – or may be transferred to the land surface by *dry deposition*, at distances far removed from their original point of release. Dry deposition involves the direct absorption of gaseous components of air into surface waters or the land surface. This movement of molecules from one phase of the environment to another needs to be considered when constructing models which attempt to predict the distribution of pollutants through different compartments (see section 3.3).

The discussion so far has been concerned with the movement of pollutants that are released into the atmosphere. Brief mention should also be made of molecules that are generated by chemical reactions in the atmosphere. As mentioned earlier, some pollutants are generated due to the interaction of chemicals released from internal combustion engines, especially under conditions that give rise to photochemical smog. Ozone is generated from molecular oxygen in the stratosphere under the influence of solar radiation. Most ozone production occurs in the equatorial zone, after which it diffuses out towards the polar regions giving rise to a more or less continuous ozone layer in the stratosphere. In the 1980s it was discovered that a hole had appeared in the ozone layer above the South pole. This was subsequently attributed to the destruction of ozone when it interacts with CFCs – volatile pollutants that can diffuse into the stratosphere.

3.3 Models for Environmental Distribution of Chemicals

Models have been constructed which attempt to describe, in mathematical terms, the movement and distribution of chemicals between different compartments of the environment (descriptive models). Sometimes they can also be used to predict the movement and distribution of environmental chemicals (predictive models). Broadly speaking, the models are either thermodynamic or kinetic. Models of the former type do not consider the dimension of time – they are concerned with the distribution that will be found when a *thermodynamic equilibrium* is reached. Kinetic models, on the other hand, are concerned with the *rates* at which processes of transfer or transformation occur. That is, they include a time factor.

For the purposes of modelling, the environment can be divided into compartments physically distinct from one another, and separated by phase boundaries (e.g. air/water, water/gas etc.). At the simplest level the distribution of a chemical between two phases *at equilibrium* is described by a partition coefficient. This represents a simple thermodynamic model where no time factor is involved. An example of this is the octanol–water coefficient (K_{Ow}).

$$K_{Ow} = \frac{\text{conc. in octanol}}{\text{conc. in water}}$$

K_{Ow} gives a measure of the *hydrophobicity* of a chemical, i.e. its tendency to move from water (a polar liquid) into a non-polar liquid which does not mix with water. True equilibrium states are usually found in closed systems where the molecule(s) under consideration are not entering or leaving the system. This is not typical of the natural environment, where the systems under consideration are usually 'open', and pollutants are entering and leaving, and undergoing chemical transformation and/or biotransformation. Here the partition coefficients will describe the distribution of a pollutant between two phases, so long as the system is in a *steady state* – where the concentrations of a pollutant in the phases under consideration are constant, and do not change with time. Such would be the case with the water of a river carrying a constant concentration of a lipophilic pollutant over a sediment high in organic matter. Some pollutant would partition into the sediment from the water, but an equivalent quantity would partition from the sediment into the water. Thus the concentration of pollutant in sediment and its ambient water would remain constant. In practice the distinction between equilibrium and steady state is not very important in the present context because, in both cases, partition coefficients effectively describe the distribution of a pollutant between two adjacent compartments.

In attempting to describe the distribution of chemicals through several compartments over relatively large areas, some success has been achieved by the use of *fugacity* models. These are, again, thermodynamic models. They are based on physicochemical properties of chemicals, which determine distribution, and on environmental variables such as temperature, pH, quality and quantity of light, and water and air movement. The environmental variables are complex, and only predictable to a very limited degree. For this reason fugacity models have had only very limited success when employed in a predictive way. However, they have been useful as evaluative models, which describe the environmental distribution of pollutants under defined conditions. One virtue of this approach is that it does give some ranking order among a group of chemicals, in regard to their tendency to move into air, sediments and other compartments of the environment.

Fugacity models to describe the distribution of environmental chemicals were first introduced by Mackay (see Bacci, 1993). The underlying principle is that fugacity is a measure of the tendency of a molecule to escape from a particular phase or compartment. It is measured in the same dimensions as pressure. When considering the distribution of a chemical through several adjoining phases, equilibrium is reached when the chemical has the same fugacity in all phases.

In any one phase, $f = C/Z$

where C = concentration of chemical in phase
Z = fugacity capacity constant
f = fugacity.

Considering now a two-phase system in equilibrium:

$$f_1 = f_2$$

where f_1 and f_2 are fugacities in phase 1 and phase 2 respectively.
Thus

$$\frac{C_1}{Z_1} = \frac{C_2}{Z_2}$$

or

$$\frac{C_1}{C_2} = \frac{Z_1}{Z_2} = K_{12}$$

where K_{12} is the partition coefficient for the chemical between phases 1 and 2. C_1 and C_2 are concentrations in phases 1 and 2, Z_1 and Z_2 are fugacity capacity constants in phases 1 and 2. Thus, the partition coefficient is the ratio of the fugacity capacities constants for the two compartments.

The distribution of a gas between air and water provides an example of how the model works. Here, the fugacity in air corresponds to the partial pressure of the gas (P_a). This is a measure of the 'escaping tendency'. The distribution of a gaseous pollutant between air and water is described by Henry's constant (H):

$$\frac{\text{concentration of gas in water } (W)}{\text{partial pressure of gas } (P)} = H$$

or

$$\frac{W}{H} = P.$$

Fugacity capacity constants (Z values) can be calculated for different compartments of the environment. The higher the Z values, the higher the expected concentrations in the compartment in question.

Fugacity models represent the environment as a number of compartments of known volume in which an equilibrium can be established. These compartments are described as 'units of world'. The compartments defined, include 'air', 'lake water', 'soil', 'sediment' and 'biota'. Biota have been subdivided into animals and plants. An example of the distribution of environmental chemicals as described by a fugacity model is shown in Figure 3.4.

It should be emphasized that these models are at best, only of limited *predictive* value, as currently used. The inability to predict various environmental

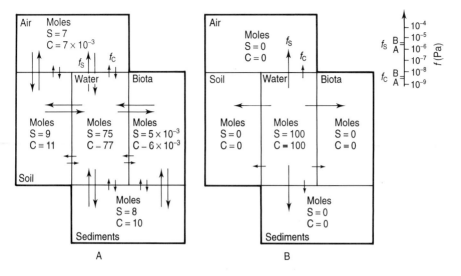

Figure 3.4 The state of the different compartments of the environment is shown at two different times. In B, the two insecticides are present only in water. In A they have moved from water into all the neighbouring compartments to achieve equilibrium. The arrows indicate direction of movement in accordance with fugacity values. Thus, in B there is only movement away from the water compartment. In A, there is equal movement in either direction on all phase boundaries because the system is in equilibrium. The number of moles (gram molecular weights) is indicated for the two insecticides. S = sulfotep; C = chlorfenvinphos. f_s and f_c refer to fugacities of sulfotep and chlorfenvinphos respectively. The long arrows are for f_s, the short ones for f_c. The major difference between the two compounds in distribution is the greater tendency for sulfotep to 'escape' from water to air. It has a higher vapour pressure (measure of fugacity). From Calamari and Vighi (1992).

parameters such as temperature and wind speed – and the fact that chemicals are seldom at equilibrium or in the steady state – has placed a limit upon their effectiveness.

Until now this discussion has been restricted to *thermodynamic* models for systems at equilibrium or in the steady state – or when approximate to one or other of these situations. *Kinetic* models have also been utilized and these are concerned with the *rates* at which processes of transfer or transformation occur. At the simplest level, the rate of transfer from one compartment to another follows 'first order' kinetics, and is described by the following equation:

$$r = -kC$$

where r is rate of transfer, k is rate constant and C is the concentration of the chemical in the phase (compartment) from which it is escaping.

Considering a pollutant that moves between compartments A and B by diffusion, a constant concentration will be reached after a certain time when equilibrium is reached. Then:

$$r_{A \to B} = r_{B \to A}$$
(rate transfer   (rate transfer
from A to B)   from B to A)

but

$$r_{A \to B} = -k_{AB} C_A$$

and

$$r_{B-A} = -k_{BA} C_B$$

where k_{AB} and k_{BA} are rate constants for movement from $A \to B$ and $B \to A$ respectively, and C_A and C_B refer to concentration in compartments A and B respectively.

It follows from this that:

$$\frac{CA}{CB} = K_{AB} = \frac{k_{BA}}{k_{AB}}$$

where K is the partition coefficient between compartment A and compartment B. In other words, the partition coefficient is the ratio of the rate constants at equilibrium. Much more complicated kinetic equations than these have been developed, but they lie outside the scope of this book.

To summarize, kinetic models can be used for environmental modelling. In theory they have an advantage over thermodynamic ones: they can be used to describe the distribution of chemicals between compartments under conditions far removed from equilibrium or steady state. However, this approach has yet to be successfully developed.

Further Reading

WAYNE, R. P. (1991) *Chemistry of Atmospheres*, 2nd Edition An authoritative and readable account of movements in the atmosphere, and of photochemical reactions.

TUREKIAN, K. K. *Oceans*, 2nd Edition Contains a description of major ocean currents.

DIX, H. M. (1981) *Environmental Pollution* Gives a wide-ranging account of the distribution and movement of pollutants in air and water.

BACCI, E. (1993) *Ecotoxicology of Organic Contaminants* Describes models for the environmental distribution of pollutants.

CALAMARI, D. and VIGHI, M. F. (1992) Describes the use of fugacity models.

4

The Fate of Metals and Radioactive Isotopes in Contaminated Ecosystems

4.1 Introduction

Four factors control the fate of inorganic pollutants in contaminated eco-systems. These are localization, persistence, bioconcentration and bio-accumulation factors and bio-availability.

4.1.1 Localization

There would be no pollution problems if all pollutants were spread evenly throughout the globe. The solution to pollution is dilution! A pollutant is toxic when its concentration exceeds a threshold value in a particular environmental 'compartment'. The ultimate compartment is the whole planet, but compartments can be individual organisms or as small as single cells, or even organelles within cells (see Figure 8.1).

At the *ecosystem* level, tall chimneys operate on the 'safe dilution' approach to discharges to the environment. For example, pollution from the nickel smelting works at Sudbury, Canada, caused severe ecological disruption to the surrounding countryside. The 'solution' was to increase the height of the chimney so that metal particulates were carried further from the factory. While the total amount of pollutants discharged remains the same, the concentration locally has markedly reduced so that plants have begun to recolonize the vicinity of the factory.

At the other end of the scale, at the cellular level, organisms may compartmentalize potential toxins in insoluble deposits to prevent interference with essential biochemical reactions in the cytoplasm. For example, the epithelium of the midgut of most invertebrates contains metal-rich granules which act as intracellular sites of storage detoxification (see Figures 8.4–8.6).

4.1.2 Persistence

Metals are non-biodegradable and do not break down in the environment. However, there is formation and degradation of specific compounds such as methyl mercury. Once metals get into soils or sediments, they have long residence times, usually of several years, before they are eluted to other compartments.

Radioactive isotopes of metals decay exponentially, and persistence is dictated by the half lives of the individual isotopes (Table 1.4). High-level waste from nuclear reactors is extremely persistent as it includes isotopes with half lives of many thousands or even millions of years.

4.1.3 Bioconcentration and Bio-accumulation Factors

Some inorganic pollutants are assimilated by organisms to a greater extent than others. This is reflected in the *bioconcentration factor* (BCF) which can be expressed in the following manner:

$$\text{BCF} = \frac{\text{concentration of the chemical in the organism}}{\text{concentration in the ambient environment}}$$

For a terrestrial organism the ambient environment is usually the soil. For an aquatic organism, it is usually the water or sediment. With inorganic chemicals, the extent of long-term *bio-accumulation* depends on the rate of excretion (see Chapter 5). Thus bio-accumulation of cadmium in animals is high relative to most other metals as it is assimilated rapidly and excreted slowly (for examples, see Chapter 11). If an organism exhibits a high bioconcentration factor for a particular substance this may be due to its biochemistry. For example, animals with a calcareous skeleton, exoskeleton or shell take up lead and/or strontium to a greater extent than those without because these two substances follow similar biochemical pathways to calcium for which the organism has evolved a high assimilation efficiency.

4.1.4 Bio-availability

Another reason for a high bioconcentration factor may be that the substance in question is more bio-available than one with a low bioconcentration factor. Methylated mercury is taken up more readily than the unmethylated form. pH has a marked effect on the solubility of metals in soils and water. If the pH declines (due for example to acid deposition) some metals become more soluble than others, and hence more bio-available. Aluminium is highly insoluble at normal to slightly acidic pH, but below about pH 4.5 its solubility increases dramatically and it becomes the most important factor responsible for fish kills in acidified lakes (see Chapter 14).

4.1.5 *'Cocktails' of Inorganic Pollutants*

The subjects of synergism and antagonism between pollutants are dealt with extensively in Chapter 10. However, it is worth discussing one aspect of mixtures of pollutants which is often overlooked, that is, the relationship between the relative toxicities of pollutants to organisms, and their relative concentrations in the field (see also the discussion of risk assessment in Chapter 6). For example, negative effects of cadmium contamination of the diet of soil invertebrates on survival, growth and reproduction can be detected at about a tenth of the concentration (by weight) at which they occur with additions of zinc to the diet. Consequently, if the concentration of zinc is ten times that of cadmium in the diet then toxicity will be due to both metals equally. However, in regions contaminated by the two metals either from past mining activity, or smelting, zinc is almost always present at about 50 times the concentration of cadmium in soils or on vegetation. Thus, zinc is responsible for toxic effects in primary consumers in these situations. Nevertheless, because cadmium has a higher bioconcentration factor in most primary consumers than zinc, predators at the next stage in the food chain may be exposed to a zinc : cadmium ratio of less than 10 and hence be poisoned by cadmium rather than zinc.

4.2 Terrestrial Ecosystems

4.2.1 *Introduction*

In terrestrial ecosystems, the soil may be contaminated with metals and radioactive isotopes as a result of previous industrial, mining or other activity, or the contamination may be due to deposition from above (Figure 4.1). Concerning the latter, this can be from agricultural practices such as the application of metal-containing pesticides, or metal-contaminated sewage sludge, or as wet or dry deposition from smelting activity, lead-containing car exhausts, atmospheric nuclear weapons testing, or accidents such as Chernobyl.

4.2.2 *Metals*

Most geological deposits of metals which became exposed at the surface due to weathering were worked out in previous centuries. This, and more recent mining activity, has left a legacy of contaminated sites in which concentrations of metals can be extremely high. Due to the long residence times of metals, mines that have been disused for many years may have a very sparse cover of vegetation (Figure 4.2A). Those plants which manage to survive are often metal-tolerant strains which are genetically distinct from their non-tolerant ancestors (see Chapter 13). Rehabilitation of such areas is difficult. Nowadays,

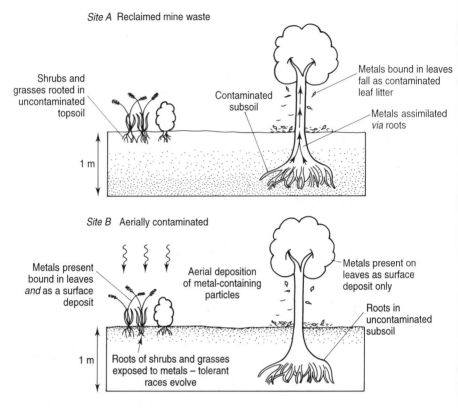

Figure 4.1 Schematic diagrams comparing the distribution of metals in (A) a disused mine site 'rehabilitated' by application of uncontaminated topsoil, and (B) a site subject to aerial contamination. Reproduced from Hopkin (1989) with permission of Elsevier Applied Science.

the most widely-used method is to 'cap' the contaminated deposit with an impermeable layer, then to cover this with top soil on which trees can be planted (Figure 4.2B). Rain falling on the soil flows over the impermeable layer to the edges of the deposit rather than through the metal-contaminated material. This approach greatly reduces the flow of metal-contaminated liquid to groundwater.

Soils may be contaminated at the surface from several sources. Before the development of synthetic organic chemicals, metal-containing pesticides were used widely. In the 19th century, it was standard practice to spray 'Bordeaux mixture' in gardens and on crops to control pests, particularly on grape vines, as the name of the mixture would suggest. Bordeaux mixture contains copper and is still widely used in the tropics as a fungicide (Lepp and Dickinson, 1994). Arsenic, lead and chromium were used also and it is still possible to detect elevated levels of these metals in garden soils of old houses.

A

B

Figure 4.2 (A) Parys Mountain, Anglesey, North Wales. During the early 19th century, this was the largest copper mine in the world. Mining ceased about 100 years ago but recolonization by vegetation has been slow due to the very high concentrations of copper in surface soils. (B) Rehabilitation of mining waste at a disused copper mine in the Gusum area, Sweden. The spoil tip is being capped with an impermeable layer prior to landscaping with a 2-metre layer of topsoil on which trees will be planted. Photographs copyright of Steve Hopkin.

One method of disposal of sewage sludge is to spread the waste on agricultural fields (the source of the term 'sewage farm'). However, because drains which 'supply' sewage treatment works also take industrial waste, concentrations of metals in the sludge can be very high. The high organic matter content of sewage has a powerful binding capacity for metals which leach very slowly down the soil profile. The number of applications of sludge that can be made to farmland is restricted by the build up of metals in soils (Alloway and Jackson, 1991).

One of the major sources of metal contamination of soils is the combustion of lead-containing petrol. In the United Kingdom, for example, leaded petrol contained about 0.4 g litre^{-1} until 1985 when the maximum permitted concentration was reduced to 0.15 g litre^{-1} (petrol in some less-developed countries still contains as much as 3 g litre^{-1} of lead). Although lead-free petrol is now available in most countries, older cars continue to emit lead to the atmosphere as fine particles. The extensive use of lead in the past has led to widespread contamination of urban soils (Culbard *et al.*, 1988). Long-range transport of the metal has occurred and elevated levels of lead can be detected in isolated regions far from industrial activity, such as Greenland (Rosman *et al.*, 1993).

The long residence time of lead in soils means that surface layers will remain contaminated with lead for several hundreds of years to come. However, the reduction in emissions to the atmosphere has been mirrored by a decline in surface deposition which, in the United Kingdom, has fallen by more than a half since 1985 (Jones *et al.*, 1991). Clear evidence of the role of cars in lead contamination has been provided following the collapse of the Berlin Wall. The influx of cars from former East Germany, which run on leaded petrol, has resulted in an increase in the lead content of the moss *Polytrichum formosum*, which has been monitored throughout the political change (Markert and Weckert, 1994).

In contrast to pollution from cars, deposition of metals from smelting activity tends to be fairly localized. One of the best studied sites in the world is the region surrounding a primary lead, zinc and cadmium primary smelting works at Avonmouth near Bristol, South West England (see Hopkin (1989) and Martin and Bullock (1994) for detailed descriptions of the area). In the close vicinity of the factory, concentrations of lead, zinc and cadmium in surface soils are at least two orders of magnitude higher than normal background levels. Significantly elevated levels of cadmium in soils can be detected up to 30 km downwind of the plant. The main effect of this heavy aerial deposition of metals is a reduction in the decomposition rate of dead vegetation which accumulates on the surface as a thick layer. Organisms such as earthworms, woodlice and millipedes, which are responsible for the initial fragmentation of leaf litter, are absent due to the metal contamination of their diet.

The mobility of metals in soils is dictated largely by the clay content, amount of organic matter and pH. In general, the higher the clay and/or organic matter content and pH, the more firmly bound are the metals, and the longer is their residence time in soil. One of the effects of acid deposition in

Europe has been 'forest dieback' which is due at least partly to nutrient deficiency (mainly magnesium). Essential elements become more mobile in acidified soils and are leached to lower soil layers to which the roots of the trees cannot penetrate (Berggren *et al.*, 1990).

In Avonmouth soils the metals exhibit the 'classic' profile of decreased concentration with depth (Figure 4.3). However, in 1976 a taller chimney was built to vent the sulphuric acid plant on the site, and the pH of soils downwind decreased due to higher acid deposition. The mobility of metals increased and

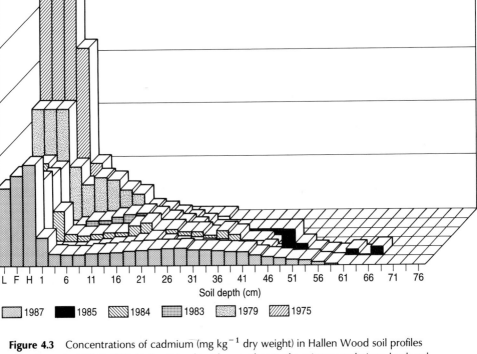

Figure 4.3 Concentrations of cadmium (mg kg^{-1} dry weight) in Hallen Wood soil profiles over the period 1975–1987. Hallen Wood is 3 km northeast of a primary cadmium, lead and zinc smelting works at Avonmouth, southwest England. Each value for the mineral soil represents analysis of a block of soil collected at depths of 0–1 cm, and then at 2.5 cm intervals to the final depth. The profiles show two main features. First, a reduction over time in concentration in the litter (L), primary (F) and secondary (H) layers. Second, a progressive wave of cadmium moving down the profile. The increased mobility of cadmium was due to increased acid deposition in the woodland following construction of a tall chimney at a sulphuric acid plant at the smelting works in the mid 1970s. Reproduced from Martin and Bullock (1994) with permission of John Wiley & Sons.

a 'progressive wave' of metals passed down through the soil profile (Figure 4.3).

4.2.3 *Radioactivity*

Contamination of soils with radioactive material is a relatively recent phenomenon since most of the elements involved did not exist naturally before the development of nuclear weapons and reactors. Some regions of the world where bombs were tested, such as the Australian and Nevada deserts, are still heavily contaminated. Any 'clean up' will have to involve removal of the surface soil, but then the problem arises of what to do with the radioactive material that is removed.

The production of nuclear energy has a good safety record relative to other methods of energy production. However, there are a number of well-publicized examples of environmental contamination. Perhaps the best known is the accident at Chernobyl complex on 26 May, 1986, when one of the reactors caught fire, eventually releasing half of its contents to the atmosphere (Edwards, 1994). Most of Europe was affected to some extent by fallout of radioactivity which was most severe to the northwest of the site. In Byelorussia (on which 70% of the total fallout from the reactor fell – Lukashev, 1993) large areas of the country are still heavily contaminated and restrictions on certain agricultural practices are still in force (Figure 4.4).

The effects of the Chernobyl fallout outside the former Soviet Union have been most persistent in Scandinavia and upland areas of Northwest Europe. The vegetation in these nutrient-poor regions is adapted to retain and recycle essential elements. Metal pollutants that are deposited from the atmosphere pass down through the soil profile extremely slowly. In Cumbria, in Northwest England, sheep on upland hill farms became contaminated with radioactive caesium for several years and were prevented from being sold for human consumption (the maximum permissible level was 1000 Bq kg^{-1}). Currently, lambs are moved to lowland pastures prior to slaughter where their radiocaesium burden is rapidly lost via faeces (Crout *et al.*, 1991). The caesium passes much more rapidly down the soil profile in these lowland fields than on the hills.

In Sweden, one of the areas most contaminated was the region occupied by the Saami community where ^{137}Cs levels in reindeer reached a mean of more than 40 000 Bq kg^{-1} (Åhman and Åhman, 1994). Since the disaster, the radioactivity of the reindeer has declined slowly and exhibited a marked seasonal fluctuation (Figure 4.5). This is correlated with the change in diet from summer to winter. During summer, reindeer feed mainly on grass, herbs and leaves which have a low radiocaesium content. During winter, lichens are an important part of the diet (up to 60%). Lichens have a radiocaesium content more than 10 times that of vascular plants from the same region. This example illustrates the importance of regular monitoring of the biota after a

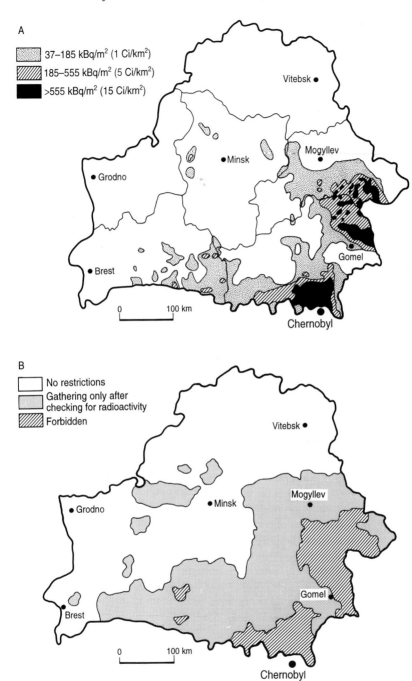

Figure 4.4 (A) Distribution of radioactive caesium in surface soils of Byelorussia after the Chernobyl accident. (B) Areas of Byelorussia with restrictions for gathering mushrooms. Reproduced from Lukashev (1993) with permission of Pergamon Press.

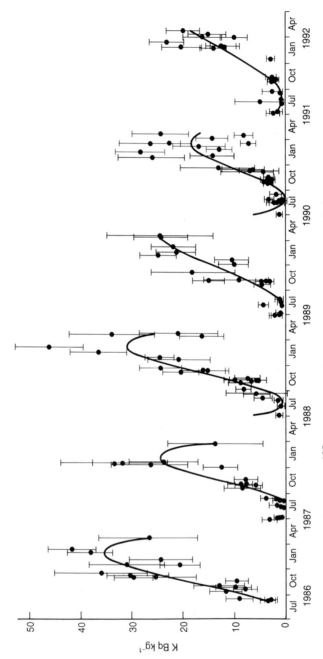

Figure 4.5 Activity concentrations of ^{137}Cs in reindeer from the Saami community, Vilhelmina Norra, Sweden from 1986 to 1992. Mean ± standard deviation from separate slaughter occasions (n = 10 to 825 animals). Reproduced from Åhman and Åhman (1994) with permission of the Health Physics Society.

pollution incident as the movement and pathways of transport may be more complicated than at first thought.

4.3   Aquatic Ecosystems

The ultimate 'sink' for metals is the ocean. However, because of the massive dilution of contaminants that occurs, it is difficult to prove that metals in the open sea are having a significant effect on the biota. Estuaries are a different story, however, and many are grossly polluted, particularly those fed by rivers that pass through heavily industrialized regions or regions of mining activity (Figure 4.6) (Grant and Middleton, 1990; Bryan and Langston, 1992). When polluted freshwater reaches the sea, the flow rate slows down, suspended sediments settle on the bottom and dissolved metals are precipitated (see Chapter 3).

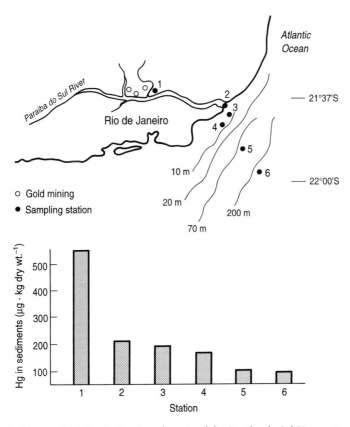

Figure 4.6   Mercury (Hg) distribution in sediments of the Paraiba do Sul River, estuary and adjacent continental shelf, Rio de Janeiro State, southeast Brazil. Reproduced from Pfeiffer *et al.* (1989) with permission of Elsevier Science Publishers.

Even if the discharges to rivers are cleaned up, the estuaries they feed may continue to be affected for many years due to remobilization of past sediment contamination. Such an effect is occurring at Minimata Bay in Japan where sediments are heavily contaminated with mercury (Figure 4.7). A new quay was built in the 1970s at which much larger ships could dock. The action of their propellors has remobilized mercury-contaminated sediment which can now be detected much further out to sea than previously (Kudo *et al.*, 1980).

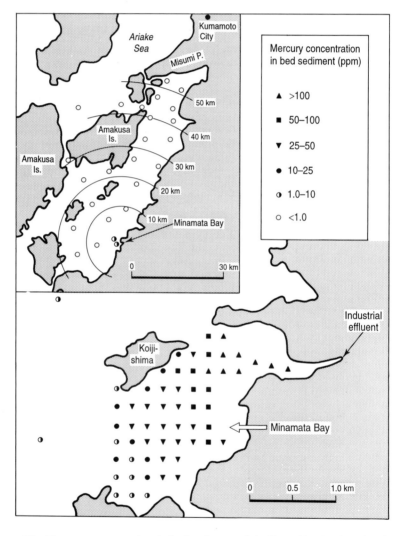

Figure 4.7 Mercury concentrations in bed sediments of the Yatsushiro Sea (1975) and Minimata Bay (1973). Reproduced from Kudo *et al.* (1980) with permission of Pergamon Press.

In water, the solubility of metals is strongly pH-dependent. Streams draining mining areas are often very acidic and contain high concentrations of dissolved metals with little aquatic life. However, as the stream becomes diluted with uncontaminated water further downstream, the pH rises and metals are precipitated onto the bed. This is the case with the stream that drains Parys mountain (Figure 4.2A) where heavy deposits of iron and copper coat the submerged rocks giving the bed of the stream a bizarre orange/brown coloration.

Acid deposition may be 'stored up' in the snow and released as a sudden 'pulse' of acidity during the spring thaw (Borg *et al.*, 1989). The resulting decline in pH of as much as one unit causes a sudden increase in the levels of soluble metals in lakes and streams (Figure 4.8).

Nowadays, the deliberate release of radioactive waste into the aquatic environment is much more tightly controlled than in the past. Until the 1980s, concrete drums containing radioactive material were routinely dumped in the ocean until the practice was banned. A bathosphere inspected some of these drums on the floor of the ocean in the early 1980s and they appeared to be intact (Sibuet *et al.*, 1985). However, their long-term integrity must remain in doubt.

The major source of radioactive contamination of the sea in Europe has been effluent from the Sellafield nuclear reprocessing plant in Cumbria, Northwest England. During the 1970s, discharges were high. Indeed, by far the most significant current problem is remobilization of this earlier contamination.

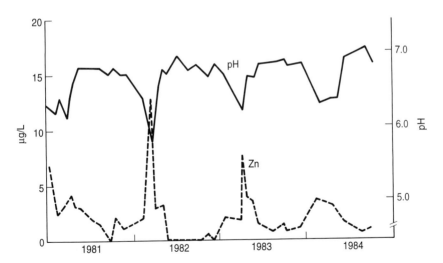

Figure 4.8 Variations with time in outlet water pH and concentrations of zinc in Lake 2011, Holmeshultasjön, Sweden. Note the close relationship between 'pulses' of lower pH in the lake and increased levels of dissolved zinc. Reproduced from Borg *et al.* (1989) with permission of Elsevier Science Publishers.

Present day discharges are very low. Much of the radioactivity is present in sediments deposited at the time which have become covered with more recent material. Thus by careful analysis of the different layers it is possible to date the 'strata' and compare their radioactivity with discharge data for the year in question (Mackenzie and Scott 1993; Mackenzie *et al.*, 1994). An example of such an approach is shown in Figure 4.9.

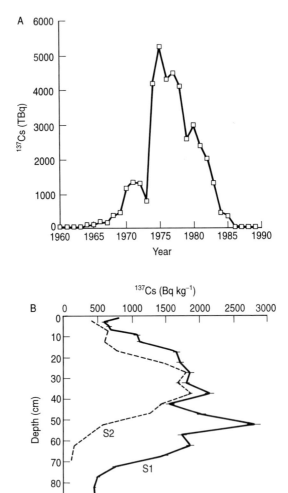

Figure 4.9 (A) Variations with time in the annual quantities of ^{137}Cs discharged from the Sellafield nuclear fuel reprocessing plant. (B) ^{137}Cs concentration profiles for Solway Firth saltmarsh sediment section S1 and S2, north of Sellafield. Reproduced from Mackenzie *et al.* (1994) with permission of Elsevier Science Ltd.

Further Reading

ALLOWAY, B. J. and JACKSON, A. P. (1991) A recent study of metal contamination of sewage sludge.

BORG, H. *et al.* (1989) A paper which clearly demonstrates the impact on metal solubility of the seasonal dip in pH in Swedish lakes.

CROUT, N. M. J. *et al.* (1991) The reasons behind the very long residence times of caesium in sheep in Northeast England and measures being taken to solve the problem.

HOPKIN, S. P. (1989) Contains general introductory chapters on metals and a detailed account of studies around the Avonmouth metal smelting works (see also MARTIN and BULLOCK (1994)).

SIMKISS K. (1993) A summary of several relevant papers on UK radioactivity after Chernobyl in this issue of the *Journal of Environmental Radioactivity*.

5

The Fate of Organic Pollutants in Individuals and in Ecosystems

The organic pollutants discussed in this book are examples of xenobiotics. A xenobiotic is here defined as a compound which is 'foreign' to a particular organism. This means that it does not play a part in normal biochemistry. By this definition, a chemical which is normal to one organism may be foreign to another. Thus, xenobiotics may be naturally occurring as well as man-made (anthropogenic), and must have existed since early in the evolutionary history of this planet. From an evolutionary point of view, the role of naturally occurring xenobiotics as 'chemical-warfare agents', is of considerable interest. For example, there is much evidence for the evolution of detoxication mechanisms by animals to give them protection against toxic xenobiotics produced by plants. Nearly all of the organic pollutants referred to in Chapter 1 are man-made xenobiotics – which do not occur in nature. It is, however, important to remember that naturally occurring xenobiotics, e.g. pyrethrins, nicotine, various mycotoxins etc., will be subject to the same toxicokinetic processes.

Toxicokinetics has relevance to ecotoxicology because it aids the understanding and prediction of the behaviour of organic pollutants within living organisms, as will shortly be explained. On the other hand, the fate of a chemical in an entire ecosystem is a more complex matter, involving movements in soils, surface waters and air, and transfer along food chains. Toxicokinetic models are sometimes valuable for prediction of the fate of chemicals in individual organisms, but more elaborate models would be required if prediction of fate in whole ecosystems were to be attempted. As discussed earlier, some success has been achieved in predicting the distribution of chemicals through major compartments of the environment (Chapter 3). However, prediction of distribution of a chemical through the different organisms constituting an ecosystem is another matter. It may well be that the whole system is too complicated to lend itself to predictive modelling of this kind.

The general principles of toxicokinetics, as they apply to lipophilic xenobiotics in individual organisms, will now be described. The emphasis will be upon animals, with some reference to plants. Toxicokinetic models will be briefly discussed making particular reference to their use for predicting bioconcentration and bio-accumulation. Finally, movement in ecosystems will be considered, dealing with terrestrial and aquatic systems separately.

5.1   Fate within Individual Organisms

5.1.1   *General Model*

The fate of a xenobiotic in an individual organism is represented in Figure 5.1. In this figure an integrated picture is given of the movements, interactions and biotransformations that occur after an organism has been exposed to a xenobiotic. The discussion that follows will focus, in the first place, on the situation that exists in animals, before drawing attention to some special features of plants. It should be stressed that this highly simplified model identifies those processes which are important from a toxicological point of view. The interplay between them will determine the toxic effect of a pollutant. For any particular chemical, interspecific differences in the operation of these processes will lead to corresponding differences in toxicity between species (selective toxicity).

The model identifies sites of uptake, metabolism, action, storage and excretion, and the arrows identify the movements of chemicals between them. The overall model will now be considered, in outline, before focusing on certain parts of it in more detail.

Once a chemical has entered an organism, four types of site are identified:

1. Sites of (toxic) action. Here the toxic form of a pollutant interacts with an endogenous macromolecule (e.g. protein or DNA) or structure (e.g. membrane) and this molecular interaction leads to the appearance of toxic

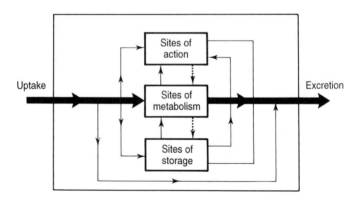

Figure 5.1   General model describing the fate of lipophilic xenobiotics in living organisms. From Walker, C. H. (Chapter 9) in Hodgson and Levi (1993).

manifestations in the whole organism. (The *chemical* acts upon the *organism*.)

2. Sites of metabolism. These are enzymes which metabolize xenobiotics. Usually metabolism causes detoxication, but in a small yet highly significant number of cases, it causes activation. (The *organism* acts upon the *chemical*.)

3. Sites of storage. Here the xenobiotic exists in an inert state from the toxicological point of view. It is not 'acting upon the organism', neither is it being 'acted upon'.

4. Sites of excretion. Excretion may be of the original pollutant, or of a biotransformation product (metabolite or conjugate). After terrestrial animals have been exposed to lipophilic xenobiotics, excretion is very largely of biotransformation products, not of original compounds (see section 5.15).

In this simple model a single box is shown for each of the categories of sites. In reality, of course, there may be more than one type of site in any particular category – and more than one location in the body for any type of site. Thus, a xenobiotic may be stored both in fat depots and in inert membranes. Also a target site for a neurotoxin (e.g. cholinesterase) may exist in both the central and the peripheral nervous system.

After uptake, pollutants are transported to different compartments of the body by blood and lymph (vertebrates) or haemolymph (insects). Movement into organs and tissues may be by diffusion across membranous barriers or, in the case of extremely lipophilic compounds, by transport with lipids. Uncharged molecules, which have a reasonable balance between oil and water solubility, tend to move across membranous barriers by passive diffusion. This happens if they are not too large (mol. wt < 800), and have an optimal octanol/water partition coefficient (K_{Ow}) for doing so (for explanation of partition coefficients see sections 3.3 and 5.1.2). Some very lipophilic compounds are transported 'dissolved' in lipoproteins. After partial degradation, fragments of lipoprotein are taken into cells such as hepatocytes by endocytosis, carrying the associated lipophilic molecules with them. Most xenobiotics are distributed throughout the different compartments of the body after uptake. Quantitative aspects of this are described in section 5.1.7.

Many of the organic pollutants discussed in this book are highly lipophilic (hydrophobic), i.e. they have high values of K_{Ow}. If not metabolized, they will be stored in fat depots, or other lipophilic sites such as membranes or lipoproteins. Such storage of potentially toxic lipophilic xenobiotics may be protective in the short term. In the long term, however, release from storage may occur, and this may lead to toxic effects in the organism. Delayed toxicity may be observed some time after initial exposure to the xenobiotic, as in the case of organochlorine insecticides such as dieldrin.

Because of their marked tendency to move into hydrophobic locations (e.g. membranes, fat depots), xenobiotics with high K_{Ow} values are not *directly* excreted in the faeces or urine of terrestrial organisms to any important extent.

Their efficient elimination is dependent upon biotransformation to water-soluble metabolites and conjugates (section 5.1.5) which are then readily excreted in faeces and/or urine. Thus the thick arrow through the middle of Figure 5.1 emphasizes the importance of this process for terrestrial animals. With aquatic organisms, however, loss by direct diffusion into the ambient water (e.g. across gills of fish) represents a very important mechanism of excretion for lipophilic xenobiotics.

The model can be subdivided into two parts. The processes of uptake, distribution and metabolism constitute the 'toxicokinetic' component. (In the case of drugs, this would be referred to as the 'pharmacokinetic component'.) Molecular interactions at the site of action are part of the toxicodynamic component (pharmacodynamic component in pharmacology). The operation of toxicokinetic processes determines how much of a toxic compound reaches the site of action (this may be the original xenobiotic or an active metabolite of same). By contrast, the nature and degree of interaction between the toxic compound and the site of action will determine the toxic response that is produced (toxicodynamic component). Sometimes it is convenient to consider these two elements separately when investigating the mechanisms that underly toxicity. A xenobiotic may be particularly toxic to a defined species for either or both of the following reasons:

1. The toxicokinetics are such that a high proportion of the active form of the xenobiotic reaches the site of action.
2. The toxicodynamics are such that a high proportion of the xenobiotic that reaches the site of action will interact there, to produce a toxic response.

Conversely, another species may be insensitive to a xenobiotic because neither of these components operates in a way that favours toxicity.

Toxicokinetic aspects of the model will now be discussed in more detail. Toxicodynamic aspects will be discussed in Part Two of the text (Chapter 7), which is concerned with effects of pollutants upon individual organisms.

5.1.2 Processes of Uptake

The most important routes of uptake are summarized in Table 5.1.

The movement of organic molecules into the organism is usually the consequence of passive diffusion across natural barriers. This is how passage across plant or insect cuticle, vertebrate skin or membranes lining gut, lungs or trachaea usually occurs. Also, very lipophilic molecules may be absorbed from the gut in association with fat (bulk transport).

The movement of organic molecules across natural barriers by passive diffusion is dependent upon them having optimal solubility properties. To move effectively across such barriers the molecules must, in the first place, have some affinity for the barrier itself, which is usually lipophilic in character. Also, they must have some affinity for the water which lies to the inside of the barrier.

Table 5.1 Major routes of uptake for organic pollutants

Type of organism	Route of uptake	Sources of pollutant
Terrestrial vertebrates	Alimentary tract	Food and ingested water
	Skin	Contaminated surfaces
		Droplets and particles in air
	Lungs	Vapour; droplets and particles in air
Terrestrial invertebrates	Alimentary tract	Food and water
	Cuticle (Insects)	Contaminated surfaces
	Body wall (Slugs, worms)	Contaminated environment e.g. soil
	Trachaea	Droplets and particles in air
Fish	Gills	Pollutants in ambient water, dissolved or suspended
	Alimentary tract	Principally food
Aquatic mammals and birds	Alimentary tract	Principally food
		Small amounts from ambient water or ingested water (birds)
Aquatic amphibia	Alimentary tract	Principally food
		Small amounts from ambient water
	Skin	Pollutants in ambient water dissolved or suspended
Aquatic invertebrates	Alimentary tract	Principally food
		Some from ambient water
	Respiratory surfaces	Pollutants in ambient water, dissolved or suspended
Plants	Leaves	Pollutants in droplets or particles[a]
		Vapours
	Roots	Pollutants dissolved in soil water[a]

[a] Important routes of uptake for herbicides and for systemic insecticides and fungicides.

Thus they should have a reasonable balance between lipid solubility and water solubility. This balance is indicated by the octanol:water partition coefficient (K_{Ow}) value (section 3.1).

Octanol is a lipophilic (hydrophobic) solvent that is immiscible with water. For efficient movement across lipophilic barriers K_{Ow} values should not be too different from 1. Values much below 1 indicate high water solubility and very low lipid solubility. Values much higher than 1 indicate very high lipid solubility (lipophilicity) but very low water solubility. The relationship between K_{Ow} and rate of movement through a lipophilic barrier is indicated in Figure 5.2.

While the K_{Ow} value gives a useful general indication of the likelihood of a molecule being efficiently taken up by passive diffusion, it must be emphasized that there is no 'optimal' K_{Ow}, which guarantees rapid uptake in *all* situations,

Rate of passive diffusion
Kow

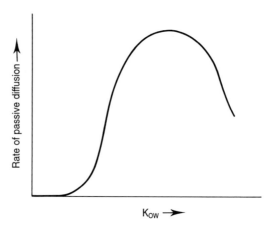

Figure 5.2 Passive diffusion of xenobiotics across a biological membrane. Movement from water through membrane into water on other side.

and other factors need to be taken into account as well. Thus, the composition and temperature of lipophilic barriers determine their state of fluidity, and therefore the ease with which molecules can diffuse into them. At low temperatures lipid bilayers can lose their fluidity, making diffusion through them difficult or impossible.

A further factor needs to be borne in mind with passive diffusion of pollutants which are weak acids or bases. Here, a state of equilibrium exists between charged and uncharged forms, which is determined by the pH of the ambient medium (Figure 5.3).

Usually, only the uncharged form will readily cross a lipophilic barrier. Thus, the uptake of weak acids is favoured by low pH, but the uptake of many weak bases is favoured by high pH. Herbicides which are weak acids (e.g. 2,4D, MCPA) penetrate plant cuticles rapidly if they are in a medium which has low pH. Within the alimentary tract of mammals, weak acids tend to be absorbed in the stomach (pH 1–2) and weak bases in the duodenum, where the pH is much higher.

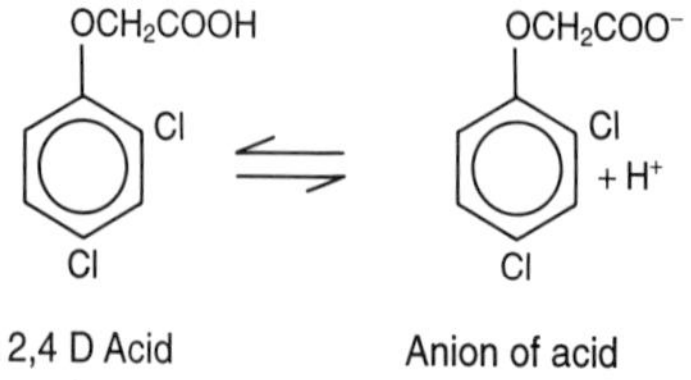

OCH2COOH
Cl
Cl
2,4 D Acid
OCH2COO−
Cl
+ H+
Cl
Anion of acid

Figure 5.3 Equilibrium of weak acid.

Returning to Table 5.1, the different types of organism will now be considered separately.

Terrestrial vertebrates and invertebrates take up lipophilic pollutants from the alimentary tract or across the skin or cuticle. Pesticides represent a very important category of pollutants in agricultural ecosystems where there can be substantial exposure to potentially toxic compounds by either or both of these routes. In general, uptake across the cuticle of insects is likely to be more important than uptake across the skin of vertebrates. This is because insects are much smaller and have much higher ratios of surface area/body volume than do vertebrates (i.e. they have much more absorbing surface per unit volume). The mobility of the organism is an important factor in determining the rate of uptake across cuticle or skin. With invertebrates, mobile predatory species will tend to come into contact with more pesticide on soil or plant surfaces than will more sedentary species. In the case of vertebrates, movement between different locations in agricultural areas will determine the extent to which they come into contact with pesticides.

Soil organisms such as earthworms, collembola and mites may be continuously exposed to persistent pesticides present in soil. This raises the issue of the availability of compounds which are bound to soil, clay and organic matter. While compounds dissolved in soil water are freely available for uptake, the availability of bound organic compounds is not well understood. The foregoing comments refer particularly to pesticides; however, similar considerations apply to other organic pollutants in terrestrial ecosystems.

Organic pollutants may also be absorbed via the respiratory system of vertebrates and invertebrates. Absorption by this route readily occurs with pollutants which are in the gaseous state. A more complex situation exists with pollutants associated with droplets or particles, which may be deposited in the respiratory tract. Examples of this include smokes and dusts emitted from factories, domestic premises or internal-combustion engines, pesticidal sprays and dusts applied to agricultural land, and rain which is contaminated with airborne pollutants. As yet, little is known about the extent to which terrestrial animals take up organic pollutants by this route.

In contrast to their terrestrial counterparts, *aquatic vertebrates and invertebrates* are exposed directly to many pollutants dissolved or suspended in surface waters. Uptake across respiratory surfaces or skin represents an important route of entry for many dissolved aquatic pollutants. However, as with soils, there is still little knowledge of the extent to which pollutants are taken up when they are bound to sediments or suspended particles.

Uptake from food may also be important in aquatic organisms. This is especially true of predatory birds, mammals and reptiles at the top of the marine food chain, which do not appear to take up pollutants directly from water to any important extent. They do not have gills, and their skins are not thought to be very permeable to organic molecules (compare the more permeable moist skins of some amphibians).

Description of the routes of uptake of pollutants would be incomplete without a brief mention of the transfer of pollutant from parents to offspring. Pollutants may be transferred across the placenta of mammals, into the developing embryo. In birds and reptiles, lipophilic compounds are transported with lipids, into the egg, and subsequently the developing embryo. Also, some lipophilic pollutants are secreted into the mammalian milk, and are thus passed to offspring during suckling.

Plants can absorb pollutants across the leaf cuticle and through the roots. These processes are well characterized for translocated herbicides, and for systemic insecticides and fungicides (Hassall, 1990). Gases can be absorbed through stomatal openings. Movement through leaf cuticle is by passive diffusion, and is dependent on the K_{Ow} of pollutants.

5.1.3 *Processes of Distribution*

In vertebrates, absorbed pollutants may travel in the blood stream and, to a lesser extent, in the lymph. Where absorption occurs from the gut, much of the absorbed pollutant will initially be taken to the liver by the hepatic portal system. Commonly, a high proportion of the circulating pollutant will then be taken into hepatocytes (first-pass effect). Entry into hepatocytes may be by diffusion across the membrane or by co-transport with lipoprotein fragments which are taken up by endocytosis. Absorption via lungs or skin may lead to a somewhat different initial pattern of distribution since the blood will travel first to tissues other than the liver. Within blood and lymph, organic molecules will be distributed between different components according to their solubility properties. Highly lipophilic compounds (high values of K_{Ow}) will be associated with lipoproteins and membranes of blood cells, with little tendency to dissolve in blood water. Conversely, more polar compounds (low values of K_{Ow}) will tend to dissolve more in water, and associate less with lipoproteins and membranes of blood cells.

Movement of organic molecules into the brain is of particular concern in toxicology since this is the site of action of many highly toxic substances (see Chapter 7). To enter the brain, organic molecules must cross the 'blood–brain' barrier. This consists, essentially, of membranes which lie between blood plasma, and the brain. In general, lipophilic compounds can cross this barrier, but charged molecules (very low K_{Ow}) cannot.

In invertebrates, movement of organic pollutants is in the haemolymph. Otherwise distribution follows a course similar to that described for mammals.

Plants transport absorbed pollutants either in the phloem (symplastic transport) or in the transpiration stream of the xylem (apoplastic transport). Molecules entering via the roots may be transported to the aerial parts of the plant in the xylem. Molecules entering via the leaf may be taken to other parts of the plant in the phloem (Hassall, 1990).

5.1.4 *Storage*

Xenobiotics may be located in positions where they are not able to interact with their sites of action, and they are not subject to metabolism. Of particular importance are lipophilic (hydrophobic) environments, especially fat depots, but also lipoprotein micelles and cell membranes which lack sites of action or enzymes that can metabolize the xenobiotic in question. (It should be emphasized that no general rule applies here – an inert membrane for one xenobiotic may contain a site of action or a detoxifying enzyme for another.) Many lipophilic xenobiotics can be stored in depot fat and some are stored as a consequence of binding to proteins. An example of the latter type of storage is the binding of rodenticide warfarin to serum albumin.

The amount of depot fat in vertebrates is subject to considerable variation. At times, when food is plentiful, fat depots may be built up – in some cases to the point where they account for 20% or more of the total body weight. An example of this is puffin (*Fratercula arctica*) chicks which are 'balls of fat' when they leave nesting burrows and find their way to the sea. At other times – when food is scarce, during illness, egg laying or migration – fat depots are run down to provide energy. The rapid mobilization of depot fat will bring a rapid release of stored pollutants into the bloodstream which will then find their way to sites of action and metabolism. Thus, in the short term, storage of lipophilic pollutants in fat depots may minimize their toxic effect. Conversely, longer term release from storage may lead to toxic effects.

The problem of delayed toxicity was well illustrated in studies on the effects of the insecticide dieldrin on Eider ducks (*Somateria mollissima*) in the Netherlands during the 1960s (Koeman and van Genderen, 1972). During the breeding season female Eider ducks died of dieldrin poisoning, males did not. At the onset of the breeding season females build up considerable fat depots, which are then utilized during the course of egg laying (large clutches of eggs are produced). In the present example, significant quantities of dieldrin were laid down in fat depots. Because of the size of the fat depots, the concentrations in other tissues were not particularly high. The dieldrin concentrations in fat were 10–20 times greater than in tissues such as liver or brain. With the mobilization of fat depots, however, blood dieldrin levels rose rapidly from about 0.02 μg ml^{-1} to 0.5 μg ml^{-1}, and birds were poisoned due to the corresponding increase of concentrations of the insecticide in the brain. Similar effects are to be anticipated in other situations where mobilization of fat depots leads to the relatively rapid release of stored lipophilic compounds of high toxicity. Other organochlorine insecticide residues (e.g. *p,p'*DDT, heptachlor epoxide), organomercury compounds, polychlorodibenzodioxins (PCDDs) and polychlorinated biphenyls (PCBs) can be redistributed in a similar fashion. Also, such compounds may cause delayed toxicity during illness, starvation or migration.

The importance of storage of persistent lipophilic compounds in depot fat is evident from many analyses of vertebrate samples from terrestrial, marine and freshwater ecosystems. A wide range of organochlorine compounds can be identified using techniques such as capillary gas chromatography. Concentrations are particularly high in predators at the top of food pyramids (see later sections of this chapter).

5.1.5 *Metabolism*

The enzymic metabolism of most lipophilic xenobiotics occurs in two phases (see Timbrell, 1992):

$$\text{Xenobiotic} \xrightarrow{\text{Phase I}} \text{Metabolite} \xrightarrow[\substack{\text{Endogenous}\\\text{compound}}]{\text{Phase II}} \text{Conjugate}$$

$$\xrightarrow{\hspace{6cm}}$$

Increasing polarity

The initial (Phase I) biotransformation involves oxidation, hydrolysis, hydration or reduction in the great majority of cases, and normally leads to the production of metabolites which contain hydroxyl groups. The hydroxyl group introduced in this initial step is necessary for most, but not all, of the subsequent conjugation reactions which constitute the second stage (Phase II) of biotransformation. These two phases lead to a progressive increase in water solubility, moving from a lipophilic xenobiotic, to a more polar metabolite and then to an even more polar conjugate. Most conjugates are negatively charged (anions), have appreciable water solubility, and are readily excreted in bile and/or urine. Following exposure of vertebrates and insects to lipophilic xenobiotics, most of the excreted products are conjugated.

The scheme shown above represents a simplification of the real situation. Phase I may involve more than one step. Also, some xenobiotics (especially if they already possess hydroxyl groups) may undergo conjugation directly. The relationship between metabolism and excretion will be explained in the next section (section 5.1.6).

In the great majority of cases, biotransformation leads to a loss of toxicity (detoxication) and is protective to the organism. However, in a small – but highly significant – number of cases, metabolism leads to an increase in toxicity (activation). In particular, oxidation in Phase I leads to the production of reactive metabolites which can bind to cellular macromolecules. Oxidation of organophosphorus insecticides such as dimethoate, diazinon, malathion, disyston, chlorpyriphos and many others leads to the production of reactive metabolites (oxons) which can phosphorylate, and thereby inhibit, acetylcholinesterase of the nervous system (Chapter 7). Oxidation of carcinogens such as benzo(a)pyrene, aflatoxin and vinyl chloride leads to the formation of reactive metabolites which can bind to DNA (Chapter 7). Thus, relatively inert molecules, which are not themselves able to cause toxic effects, are converted to reactive metabolites having very short biological half lives, which can cause cellular damage. One of the curious features of biochemical toxicology is that many of the most destructive types of molecules are reactive metabolites which are difficult or impossible to detect because of their short biological half lives. Proof of their existence and the damage that they cause often depends on identification of the modifications that they cause to cellular macromolecules – e.g. inhibited acetylcholinesterase or damaged DNA (see Biomarkers, Chapter 9).

The remainder of this section will be devoted to a brief description of the major enzymes concerned with biotransformation, using organic pollutants as examples.

The major classes of enzyme which metabolize xenobiotics in vertebrates and invertebrates are presented in Table 5.2. Many of the enzymes responsible for Phase I biotransformations are located in the endoplasmic reticulum – notably that of the liver (vertebrates), hepatopancreas, fat body and gut (invertebrates). Lipophilic xenobiotics tend to move into the endoplasmic reticulum, but their more polar biotransformation products tend to partition out into the cytosol. Conjugating enzymes such as sulphotransferases and glutathione-S-transferases located in the cytosol can then conjugate the metabolites. The conjugates are readily excreted in urine and/or bile. Thus, the increase in polarity which results from the sequential biotransformation shown above, leads to movement first from membrane to cytosol, and then from cytosol to urine and/or bile. Although many of the types of enzymes shown in Table 5.2 are also found in plants, the activities are usually low in comparison to those found in animals.

Of the enzymes responsible for Phase I biotransformations, the *microsomal monooxygenases* (*mixed-function oxidases*), are the most versatile. They are found in all vertebrates and invertebrates, and are present in the endoplasmic reticulum of a variety of tissues. Low levels have been found in plants. Unlike other Phase I enzymes, the biotransformations that they catalyze are not restricted to any particular functional group. They are able to metabolize the great majority of lipophilic xenobiotics – so long as they are not too large (large molecules would not be able to fit into the available space adjacent to the catalytic site). The main exception to this rule is presented by highly halogenated compounds such as certain PCBs and PBBs – microsomal monooxygenase does not work effectively against C–halogen bounds (i.e. C–Cl, C–Br, etc.).

Oxidations by the microsomal monooxygenase system depend upon the activation of molecular oxygen (O_2), after it has been bound to an associated haemprotein, cytochrome P_{450}. Activation is accomplished by the transfer of electrons to bound O_2, which then splits, one atom being used to oxidize the substrate (e.g. organic pollutant), the other being used to form water. Electrons for this purpose come from NADPH (sometimes from NADH). The overall

Table 5.2 Enzymes that metabolize lipophilic xenobiotics (Phase I)

Name	Principal location	Cofactors	Substrates
Microsomal monooxygenases (mixed-function oxidases)	Endoplasmic reticulum of many animal tissues, especially liver of vertebrates, hepatopancreas, fat body and gut of invertebrates	NADPH (NADH) O_2	Most lipophilic xenobiotics of molecular weight <800
Carboxyl esterases	Endoplasmic reticulum of many animal tissues; also found in cytosol and in serum/plasma of vertebrates	None known	Lipophilic carboxyl esters
'A' esterases	Endoplasmic reticulum of certain cell types of vertebrates; mammalian serum/plasma (associated with high density lipoprotein)	Ca^{2+}	Organophosphate esters
Epoxide hydrolases	Principally endoplasmic reticulum of animal cells; some in cytosol	None known	Organic epoxides
Reductases (a number of different enzymes can express this activity at low levels of dissolved oxygen, including certain flavoproteins and haemproteins)	Endoplasmic reticulum and cytosol of a number of types of animal cell	NADH/NADPH	Organonitrocompounds, some organohalogen compounds (e.g. p,p'DDT)

reaction can be written thus:

$$NADPH_2 + X + O_2 \rightarrow XO + H_2O + NADP$$

where X = substrate.

The haemprotein cytochrome P_{450} exists in many different forms which have contrasting, yet overlapping, substrate specificities. Some forms of cytochrome P_{450} are readily *inducible*. The appearance of a lipophilic xenobiotic inside the body can trigger the synthesis of more cytochrome P_{450}, leading to more rapid biotransformation of xenobiotics. This is normally protective, enabling the organism to eliminate more rapidly the xenobiotic which originally caused induction. (There are, however, cases where the chemical causing induction is not metabolized by the cytochrome P_{450} that is induced, e.g. highly chlorinated PCBs.)

Although oxidation by cytochrome P_{450} causes detoxication in the great majority of cases, often introducing hydroxyl groups to facilitate conjugation, there are some very important exceptions to this rule. The oxidative desulphuration of organophosphorus insecticides which possess a thion group (e.g. diazinon, dimethoate, disyston and malathion) leads to the formation of oxons which are active anticholinesterases (see Figure 5.4). Some organochlorine insecticides of the cyclodiene group are converted to stable and toxic epoxides (Figure 5.4). Thus, aldrin is converted to dieldrin, and heptachlor to heptachlor epoxide. Also, a number of carcinogens are activated by the same system. Polycyclic aromatic hydrocarbons such as benzo(a)pyrene (Figure 5.4) and aflatoxin B are converted into epoxides which are strongly electrophilic, and can bind to DNA. Nitrosamines are converted to methyl radicals and other reactive species when oxidized by cytochrome P_{450}. Vinyl chloride is also activated by cytochrome P_{450}.

Hepatic microsomal monooxygenase (HMO) activity towards xenobiotic substrates varies substantially between different groups, species and strains. In omnivorous and herbivorous mammals, activity is inversely related to body weight (Figure 5.6), small mammals tending to have more activity per unit body weight than large mammals. Fish have relatively low HMO activities, with no clear relationship to body weight. Birds have variable activities. Omnivorous and herbivorous species have HMO activities comparable to mammals of similar body size (on average a little lower), but fish-eating birds and specialized predators (e.g. sparrowhawk, *Accipiter nisus*) have activities comparable to fish (Walker, 1980; Ronis and Walker, 1989). However, birds generally, and fish-eating birds in particular, show the same relationships between HMO activity and body weight observed in mammals.

These differences are explicable in terms of detoxifying function of HMO. Terrestrial vertebrates depend upon HMO for the effective elimination of lipophilic xenobiotics but fish have far less dependence on metabolic detoxication because they 'excrete' uncharged lipophilic molecules into ambient water by passive diffusion. (The same argument can be applied to amphibia which lose

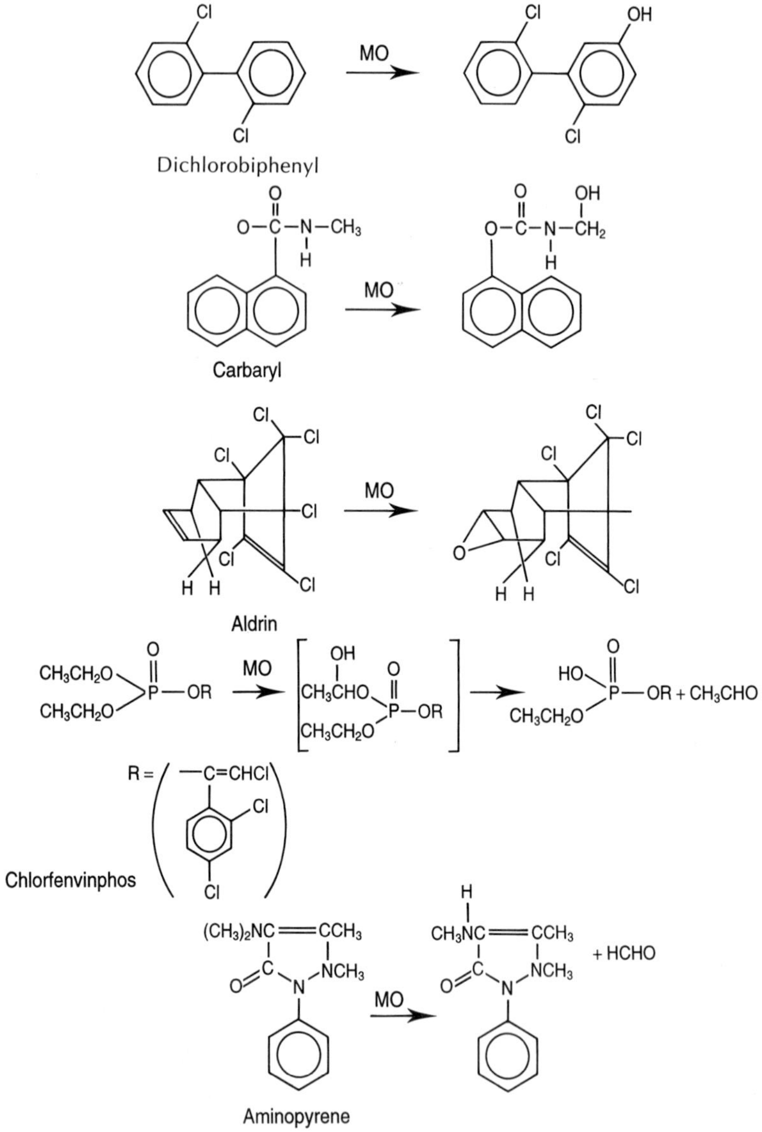

Dichlorobiphenyl

Carbaryl

Aldrin

Chlorfenvinphos

Aminopyrene

Figure 5.4   (i) Phase I biotransformations. MO = microsomal monooxygenase.

xenobiotics by diffusion across skin.) Small mammals have much higher surface area/body volume ratios than large mammals. Consequently they take in food – and associated xenobiotics – much more rapidly, in order to obtain sufficient metabolic energy to maintain body temperature. Thus they need higher levels of detoxifying enzymes than large mammals, because they take in xenobiotics more rapidly. This argument does not apply to poikilotherms, so it is not surprising that the small amount of HMO possessed by fish is not

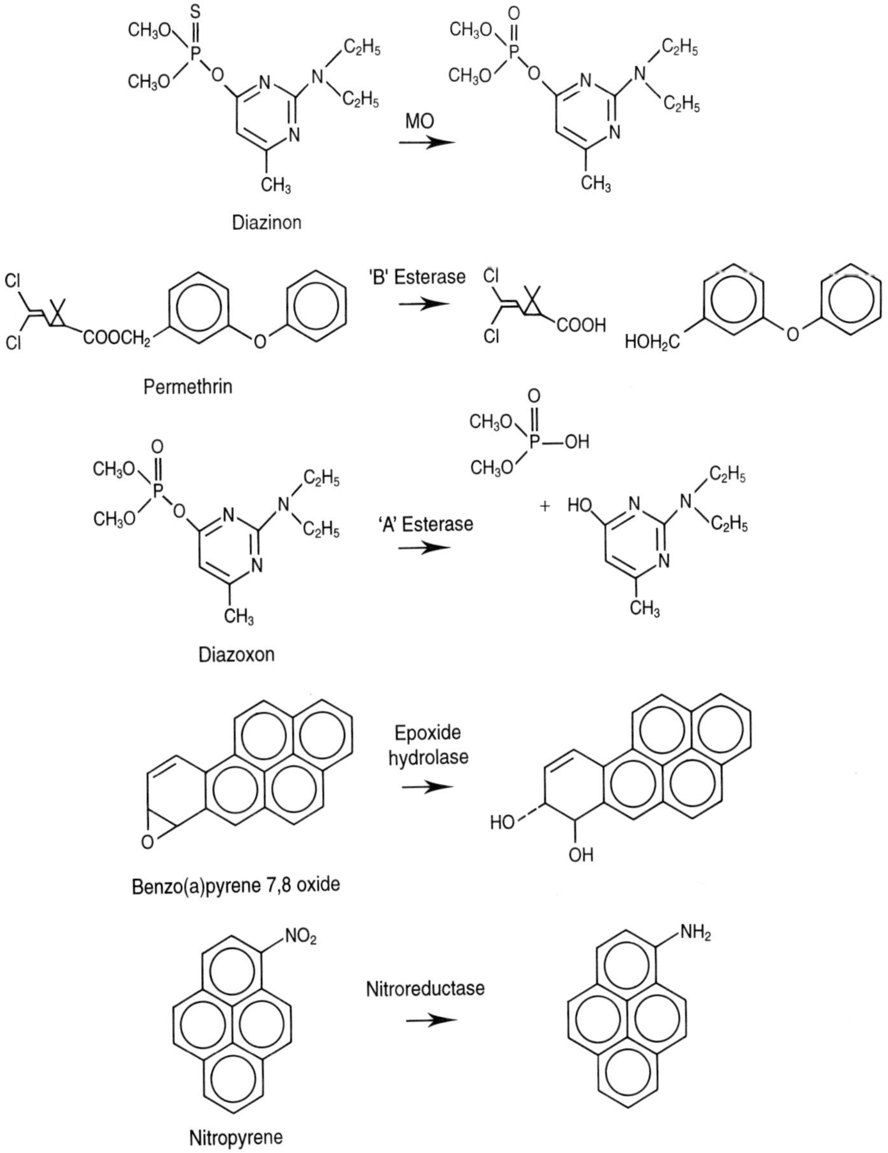

Figure 5.4 (ii) (*Continued*).

obviously related to body size – in contrast to the situation in mammals and birds. With specialized predators such as fish-eating birds and raptors like the sparrowhawk (very largely bird-eating), the need for detoxication is small, because their food does not contain many xenobiotics. Indeed the food is similar in composition to the predators themselves (especially with bird-eating

81

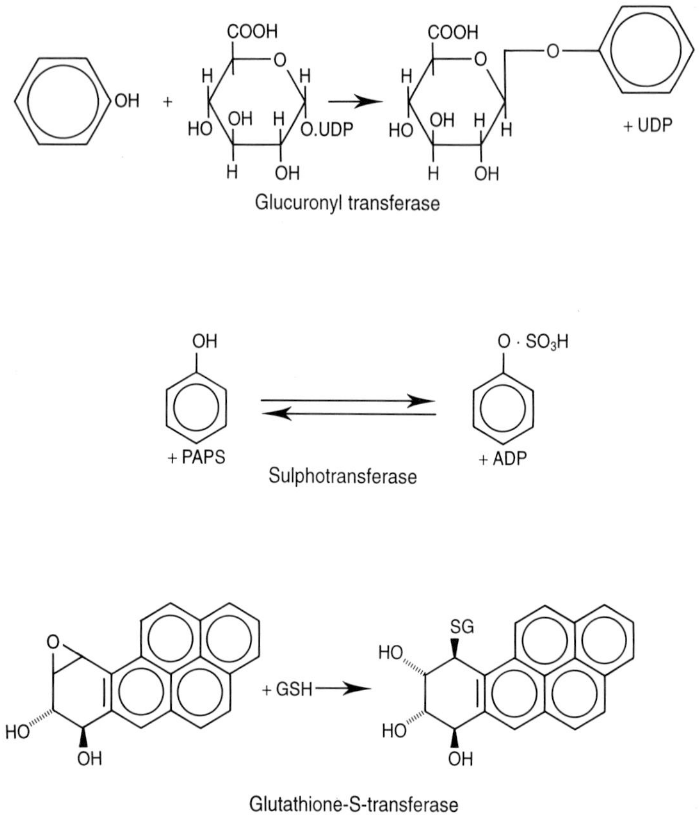

Figure 5.5   Phase II biotransformations.

predators like the sparrowhawk and peregrine falcon). Plants, by contrast, contain many compounds which are xenobiotics to animals. Some of these compounds have the function of protecting plants against 'grazing' by animals. Thus, the variations in HMO activity shown in Figure 5.6 can be explained in terms of the requirements of different species for detoxication mechanisms against liposoluble xenobiotics.

Turning now to *esterases*: Aldridge (1953) distinguished between 'A' esterases which hydrolyze organophosphates, and 'B' esterases which are inhibited by them.

Carboxyl esterases are examples of 'B' esterases and constitute an important group of detoxifying enzymes, which hydrolyze lipophilic carboxyl esters to form acids and alcohols (Figure 5.4). They are widely distributed in nature, being found in membranes (especially endoplasmic reticulum) from a variety of tissues in all animals so far investigated. They are also found in cytosol and in vertebrate plasma/serum. A number of different forms have been recognized, showing contrasting yet overlapping substrate specificities. Purified carboxyl

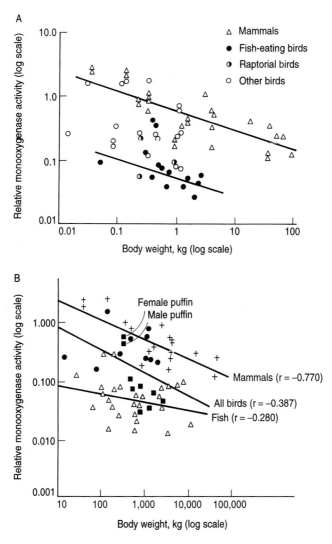

Figure 5.6 Monooxygenase activities of mammals, birds and fish. (A) Mammals and birds from Ronis and Walker (1989) (B) Mammals, birds and fish From Moriarty and Walker (1987). Activities are of hepatic microsomal monooxygenases to a range of substrates, expressed in relation to body weight. Each point represents one species (males and females are sometimes entered separately).

esterases from mammalian liver endoplasmic reticulum metabolize both xeno-biotic esters and endogenous ones. In mammals, the range of different forms of carboxyl esterases differs considerably between tissues. Liver, for example, has a wide range of forms, but blood and muscle have a smaller number of forms.

'A' esterases are enzymes which can hydrolyze organophosphorus triesters and diesters which possess an oxon group (P=O) (Figure 5.4(ii)). These esters are lipophilic in character and this type of esterase does not appear to hydrolyze organophosphates which are ionized. The known substrates are mainly organophosphorus insecticides, although some forms of 'A' esterase can hydrolyze nerve gases (e.g. soman and tabun). In vertebrates, 'A' esterase activity is found in the endoplasmic reticulum of liver and other tissues. In mammals, activity is also found in serum/plasma, some of it in association with high density lipoprotein. 'A' esterase exists in a number of different forms. Those forms which hydrolyze organophosphorus insecticides, are calcium-dependent, and lose their activity in the presence of chelating agents which bind calcium.

There are some marked species differences in regard to 'A' esterases. In contrast to mammals, birds have little or no serum/plasma 'A' esterase (Walker and Thompson, 1991; Walker *et al.*, 1991a). Some species of insects have no 'A' esterase activity.

Epoxide hydrolases are found, principally, in the endoplasmic reticulum of animals, mammalian liver being a particularly rich source. A different form of the enzyme is found in cytosol. Epoxide hydrolases hydrate a wide range of aromatic and aliphatic epoxides to form trans diols and do not require co-factors (Figure 5.5). Epoxide hydrolases of the endoplasmic reticulum hydrate epoxides generated by microsomal monooxygenases. In some cases this represents a protective function, because the epoxides are strong electrophiles which can form adducts with cellular macromolecules (see Chapter 7). Epoxide hydrolases can metabolize endogenous as well as xenobiotic substrates. Epoxides of steroids and of insect juvenile hormone are examples of endogenous substrates. As with monooxygenases and esterases, this enzyme generates metabolites with hydroxyl groups, which are then available for conjugation.

Reductases are enzymes which can catalyze the transfer of electrons to organic molecules such as nitroaromatic and organohalogen compounds (Figure 5.4). Whether enzymes operate as reductases or not often depends upon the availability of oxygen. Where oxygen is freely available this can act as an electron acceptor (i.e. electrons will flow to oxygen, rather than to the organic molecule). Both flavoproteins and haemproteins (e.g. cytochrome P_{450}) have been shown to act as reductases when oxygen levels are low. The flow of electrons to oxygen leads to the formation of oxygen radicals (e.g. superoxide anion and hydroxyl radical) which can cause cellular damage. This question will be discussed later in this section. When nitroaromatic compounds are reduced, they are converted to amines. Reduction of halogenated compounds leads to dehalogenation (e.g. conversion of p,p'DDT to p,p'DDD).

Of the Phase II enzymes which are responsible for conjugation, *glucuronyl transferases* are largely confined to membranes – especially the endoplasmic reticulum of vertebrate liver. These enzymes catalyze the interaction between lipophilic organic molecules (xenobiotic and endogenous) with labile hydrogen atoms (usually of hydroxyl groups) and glucuronic acid to form glucuronides (see Figure 5.5). The glucuronic acid is supplied by a nucleotide cofactor – uridine diphosphate glucuronic acid (UDPGA), which is synthesized in the cytosol. Thus, the glucuronyl transferase enzyme and its associated lipophilic substrate receive the polar cofactor from the hydrophilic environment of the cytosol. Glucuronides exist very largely as anions at cellular pH (approximately 7.4), and move away from the membrane and into the cytosol when they are formed (Table 5.3).

Glucuronides are subject to hydrolysis by enzymes termed glucuronidases, (e.g. in the alimentary tract) – thus leading to a reversal of the reaction shown in Figure 5.5, and a release of the hydroxyl compound that was originally conjugated. The consequences of this will be discussed later when considering excretion. Glucuronyl transferases, like other enzymes concerned with xenobiotic metabolism, exist in a number of different forms.

Sulphotransferases exist in the cytosol of various cell types (especially liver) in a number of different forms. They catalyze the transfer of the sulphate group from phosphoadenine phosphosulphate (PAPS) to xenobiotic and endogenous substrates that possess a free hydroxyl group (Figure 5.5). The cofactor PAPS is generated in the cytosol. The resulting sulphate conjugates exist as anions with appreciable water solubility. Like glucuronides, sulphate conjugates are subject to enzymic hydrolysis – in this case catalyzed by sulphatases.

Glutathione-S-transferases are found in the cytosol of many cell types, especially vertebrate liver. A number of different forms are known in vertebrates and invertebrates. These enzymes catalyze the conjugation of reduced glutathione to a variety of xenobiotics which are of electrophilic character (Figure 5.5). This conjugation would proceed naturally, in the absence of the enzyme, but very slowly. The enzyme speeds up the reaction by binding the reduced glutathione in close proximity to the xenobiotic.

Conjugations with glutathione are not dependent upon the presence of hydroxyl groups. Important substrates include certain organohalogen compounds and organophosphorus insecticides. The glutathione conjugates so formed often undergo further modification before excretion. In vertebrates, further metabolism of the glutathione moiety leads to the formation of mercapturic acid conjugates, which are usually the predominant excreted forms. Both glutathione conjugates, and the mercapturic acids derived from them, are anionic in character.

Conjugation in Phase II metabolism promotes excretion, and with very few exceptions, has a detoxifying function (however, see section 5.1.6 for further discussion). The conjugating enzymes discussed here are very important in vertebrates and invertebrates. However, a number of other types of conjugates are known, which are often 'group specific'. This is particularly true of peptide conjugation.

Before leaving the subject of metabolism, mention should be made of oxyradical formation. As discussed earlier when considering reductases, electrons

Table 5.3 Enzymes that metabolize xenobiotics (Phase II)

Name	Principal location	Cofactors	Substrates
Glucuronyl transferases	Endoplasmic reticulum of many animal cells	UDP-glucuronic acid	Especially organic compounds with free OH groups; also some organic compounds with free -SH or NH_2
Sulphotransferases	Cytosol of many animal cells	Phosphoadenine phosphosulphate	Many organic compounds with free OH groups
Glutathione-S-transferases	Mainly in cytosol of many types of animal cell; sometimes a little in endoplasmic reticulum	Reduced glutathione	Many foreign electrophiles including some organohalogen compounds and many organic epoxides

can be passed on to molecular oxygen, causing the generation of highly reactive oxyradicals. Examples include the superoxide anion ($O_2^{\cdot-}$), the hydroxyl radical ($OH^{\cdot}$) and hydrogen peroxide (H_2O_2). A description of the routes by which these are formed lies outside of the scope of this book. The significant point, in the context of ecotoxicology, is that pollutants may promote the formation of such oxyradicals, and that this may lead to cellular damage. Thus some organic pollutants (e.g. paraquat herbicide and nitro-aromatic compounds) may pass electrons on to oxygen to form oxyradicals. Additionally, certain compounds may divert the flow of electrons to oxygen (e.g. PCBs which combine with cytochrome P_{450}, but are not themselves metabolized).

Cells possess enzyme systems which detoxify oxyradicals – e.g. superoxide dismutase, catalase and peroxidase. However, they are not necessarily able to cope with increased rates of formation of oxyradicals caused by the action of organic pollutants. This is an important but difficult area of ecotoxicology, about which far too little is known.

5.1.6 Sites of Excretion

The excretion of xenobiotics and their metabolites and conjugates has been studied in some detail in vertebrates. Far less is known about these processes in invertebrates, insects having received more attention than other groups. This account will be mainly concerned with the situation in vertebrates with some references to invertebrates.

Many aquatic vertebrates can 'excrete' lipophilic xenobiotics by diffusion into ambient water. Fish can do this across gills, amphibia such as frogs, across permeable skin. Aquatic birds, on the other hand, do not have permeable membranes in contact with water; skin and feathers appear not to be readily permeable to pollutants. With aquatic mammals (whales, porpoises and seals), the skin also seems to be relatively impermeable to such compounds.

With terrestrial vertebrates and invertebrates, the effective elimination of lipophilic xenobiotics depends upon converting them into water-soluble metabolites and conjugates which can be excreted (some highly lipophilic compounds are excreted to a limited extent in milk (mammals) or eggs (birds and reptiles and invertebrates)). Vertebrates excrete these biotransformation products in bile and/or urine.

Considering excretion in urine first; soluble metabolites and conjugates are removed from blood in the glomerular filtrate, which then passes down the proximal and distal tubules of the kidney. During the movement of the filtrate down the renal tubules, some xenobiotics pass into the tubular lumen from the plasma by passive diffusion. Also, there may be some active transport of weak acids and bases across the walls of the proximal tubules into the lumen. There may also be some re-absorption of xenobiotics from the tubular lumen, into the plasma. Eventually urine collects in the bladder and is voided independently of the faeces in the case of mammals, but is combined with faeces in the case of birds, reptiles and amphibia.

Conjugates and metabolites may also be excreted via bile. In the first place these move across the plasma membrane of the hepatocytes into bile canaliculi (Figure 5.7). Bile then passes into the main bile duct and is eventually released into the duodenum. Once in the duodenum, conjugates may pass completely through the alimentary tract to be voided with faeces. Since conjugates are usually polar (in most cases they are anions), they are not readily reabsorbed by passive diffusion across the wall of the alimentary tract. If, on the other hand, they are broken down in the gut, e.g. by the action of enzymes such as glucuronidases and sulphatases, the metabolites (or, sometimes, original xenobiotics) so released, may be re-absorbed into the blood stream by passive diffusion. This is because they are uncharged, and can therefore readily cross membrane barriers. The re-absorbed compounds are then returned to the liver, reconjugated and the cycle repeated. This process is termed 'enterohepatic circulation'. Some xenobiotic metabolites may be recycled many times before they finally appear in faeces. The process of enterohepatic circulation may have toxicological consequences. Recycled metabolites may have toxic effects. Also metabolites which are themselves of low toxicity, may be transformed into toxic secondary metabolites in the gut (e.g. by microbial action). In mammals, excretion via urine does not raise the problem of enterohepatic circulation, or of further biotransformation within the gut.

The extent to which excretion occurs in bile or in urine depends upon the molecular weight of the xenobiotic and the species in question. As noted above, the majority of excreted conjugates are organic anions. Where their molecular weights are below 300, excretion is predominantly in the urine. Above molecular weight 600, it is predominantly in the bile. Between these

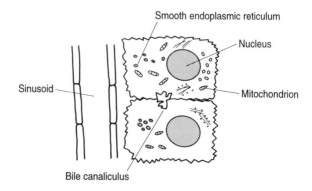

Figure 5.7 Excretion in bile – transverse section of liver cell showing bile canaliculus.

limits, the preferred rate of excretion depends upon the species and upon the molecular weight of the anionic conjugate (Figure 5.8). Threshold molecular weights have been proposed for anionic conjugates in different species (Figure 5.8). These are weights above which there is likely to be appreciable ($<10\%$) of total excretion in bile. The following figures have been proposed: rat 325 (± 50), guinea pig 425(± 50) and rabbit 475(± 50). Thus rats show a greater tendency than rabbits to excrete via bile, and may, therefore, be more susceptible to the toxic action of certain compounds for the reasons stated above.

The situation in terrestrial invertebrates appears to be similar to that for vertebrates except that the hepatopancreas or fat body serves the function of the liver.

5.1.7 Toxicokinetic Models

The processes by which pollutants are taken up, distributed, stored, metabolized and excreted have already been discussed, and are summarized in Figure 5.1. From a toxicological point of view, a living organism may be divided into different sites, and these may be ascribed different functions – uptake, storage, action, metabolism and excretion. The model described earlier is purely descriptive and not quantitative. It does not define the rates at which the processes of transfer or biotransformation occur. In toxicokinetics, the central interest is in the rates of these processes, the determination of which can lead to the development of both descriptive and predictive models. In toxicokinetic

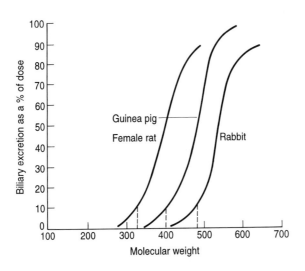

Figure 5.8 Excretion routes of anionic conjugates. From Moriarty (1975).

terms, the organism can be divided into compartments which usually represent particular organs and tissues. Each of these compartments contains a discrete quantity of xenobiotic which is subject to particular rates of transfer and bio-transformation. These compartments may or may not correspond to the sites defined in Figure 5.1.

The development of multicompartmental models represents an ideal which is seldom realized in practice. While it is desirable to have maximum information about the kinetics of particular xenobiotics in those compartments of the body that are of toxicological interest, a large amount of work is involved even for one pollutant and one species. In ecotoxicology such an approach is clearly of little value, because concern is about a large range of organisms and organic pollutants. Interest is largely restricted to simple models which treat the whole organism as a simple compartment. The following account will be concerned principally with these.

If an organism is continuously exposed to a constant level of an organic pollutant, the concentration of the pollutant in the whole organism will increase over a period of time (Figure 5.9). An aquatic organism may be exposed to a pollutant dissolved in ambient water; a terrestrial organism to a pollutant in its food. Initially the rate of increase in the tissue concentration of the pollutant will be rapid, but will then begin to tail off, and will eventually reach a plateau, so long as a lethal concentration is not reached beforehand. When the pollutant concentration reaches a plateau, the system is said to be in a 'steady-state'. The rate at which the pollutant is being taken up is balanced by the rate at which it is being lost. (Loss may be due to metabolism and/or to direct excretion of the original pollutant.) If exposure to the pollutant now ceases, the tissue concentration of the pollutant will begin to fall. In the simplest situation, where the kinetics of loss can be described by a single rate constant (one compartment model), the rate of loss is proportional to tissue concentration. First-order kinetics apply and this situation can be described

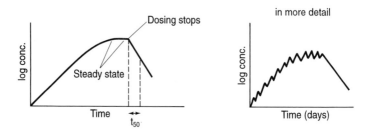

Figure 5.9   Kinetics of uptake and loss. When an animal is continually dosed with a chemical, the log conc. increases with time until a steady state is reached. When dosing of the chemical stops, the log conc. falls linearly with time if first-order kinetics apply (as in a one-compartment model).

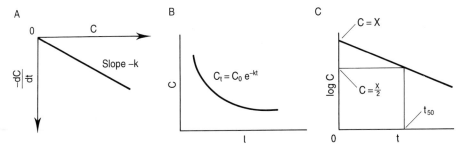

Figure 5.10 Kinetics of loss. C = tissue concentration; t = time (minutes or hours); t_{50} = half life; x = initial tissue concentration. (A) Rate of loss of chemical is proportional to tissue concentration. (B) The tissue concentration of chemical falls exponentially with time. (C) The log of the tissue concentration of a pollutant falls linearly with time.

by the equation:

$$-\frac{dC}{dt} = KC \text{ or } C_t = C_0 e^{-Kt}$$

where C is the concentration in whole organism, C_0 is the concentration at the time when exposure ceases, C_t is the concentration at time t, K is the rate constant for loss and $-dC/dt$ is the rate of loss of pollutant from the organism. This negative exponential decline in concentration is illustrated in Figure 5.10B.

It follows that the log concentration of the xenobiotic (log C) is linear with respect to time (Figure 5.10C). The biological half life (t_{50}) is the time that it will take for this concentration to fall by one half, and this can be readily determined from this plot of log C versus time.

Sometimes the whole organism does not behave as a single compartment, in which case more complex equations are needed to describe the rate of loss after cessation of exposure. Sometimes the rate of loss is biphasic, and here the loss can be described by a more complex equation derived from a two-compartment model.

Figure 5.9 represents a simplified one-compartment situation, which assumes that the rate of uptake of xenobiotic is constant, and that the state of the organism remains the same. The organism is continuously exposed to a xenobiotic until a steady state is reached, after which dosing is discontinued. In practice, the rate of uptake is not absolutely constant. For example, if the source of the xenobiotic is food, there is likely to be diurnal variation in the rate of ingestion pattern. Sometimes the activity of enzymes will change due to induction or inhibition, which thus can change the rate at which a xenobiotic is lost – and ultimately the steady state concentration for any defined rate of uptake of xenobiotic. In reality, the change of xenobiotic concentration with time is usually more complex than the simple situation shown in Figure 5.9.

5.1.8 *Toxicokinetic Models for Bioconcentration and Bio-accumulation*

Toxicokinetic models have been developed which describe – and most importantly, predict – the degree of bioconcentration or bio-accumulation of organic pollutants by animals continuously exposed to xenobiotics, when in the *steady state* (Moriarty, 1975; Moriarty and Walker, 1987; Moriarty 1988; Walker, 1990b). It is important to emphasize the value of steady state models. The tissue concentrations in the steady state represent the maximum levels that are to be expected, given a particular level of exposure. Further, they are not time-dependent. Bioconcentration or bio-accumulation factors determined before the steady state is reached are of little value, since they only apply to particular periods of exposure, and do not indicate the maximum concentration that may be reached.

In this account, the following definitions will be used:

$$\text{Bioconcentration factor (BCF)} = \frac{\text{concentration in organism}}{\text{concentration in ambient medium}}$$

Bioaccumulation factor (BAF)

$$= \frac{\text{concentration in organism}}{\text{concentration in its food (or ingested water)}}$$

For reasons stated above, these factors should be determined when the system is in the steady state. Bioconcentration factors are important in aquatic ecotoxicology, where the ambient medium is a major source of organic pollutants. Bio-accumulation factors are critical in terrestrial ecotoxicology, where food (or, in some cases, ingested water) represents a major source of organic pollutants.

For aquatic organisms (e.g. the edible mussel, *Mytilus edulis*), a close relationship has often been demonstrated between bioconcentration factors for lipophilic organic pollutants, and their K_{Ow} values (Figure 5.11). Here, a simple situation applies. Most of the uptake and loss of xenobiotic is accomplished by passive diffusion, with metabolism playing only a minor role

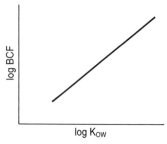

Figure 5.11 Bioconcentration factors and K_{Ow} values.

(exchange diffusion). At the steady state, the BCF is determined by the partition between water and the hydrophobic components of the organism. In the case of non-polar xenobiotics of high lipophilicity, the partition lies very much in the direction of the organism. In other words, high BCF values are associated with high values for K_{Ow}. (Sometimes K_{Ow} values are transformed to pK_{Ow} values, by converting them to log_{10} values.)

In more complex situations – where other sources of uptake and loss become more important – this simple model will break down. Thus, with fish, BCF values for certain pollutants fall well below the values predicted by a plot such as that shown in Figure 5.11 because they are rapidly metabolized, i.e. they are less than would be expected from K_{Ow} values. The rate of loss is then greater than would be expected from passive diffusion alone.

For terrestrial organisms a different approach is required to predict BAFs. In the simplest situation both uptake and loss may be ascribed to single processes, each of which will be governed by one rate constant (Figure 5.9). With strongly lipophilic compounds, food may represent the main source of pollutant, while metabolism may represent the main source of loss (Table 5.4) (Walker 1987, 1990b). Here it should be possible to develop predictive models for bio-accumulation in the steady state, using rate constants for these two processes, so long as metabolism is simple. As yet, however, no such models have been validated.

There is much interest in predictive models which will enable reasonable assessments to be made of the extent to which various organic pollutants may be bioconcentrated or bio-accumulated by different organisms. Such models could be of great assistance in the process of risk assessment for environmental chemicals (Chapter 6). They would need to be simple and cheap to operate. The major routes of uptake and loss of organic pollutants by organisms from

Table 5.4 Model system for bio-accumulation of lipophilic pollutants[a,b]

Type of organism	Routes of uptake			Routes of loss	
	Diffusion	Food	Water	Diffusion	Metabolism
Aquatic					
Molluscs	+ + + + +	(+)		+ + + + +	
Fish with substantial enzyme activity	+ + + +	+ → + + +		+ + + +	+ → + + +
Terrestrial					
e.g. predators		+ + + +	(+)		+ + + + +

[a] The importance of routes of uptake and loss are indicated by a scoring system on the scale + to + + + + +.
[b] After Walker (1975).

terrestrial and aquatic habitats are summarized in Table 5.4, and these need to be taken into account in the development of predictive models. To date, the only model which meets these criteria is the one which uses K_{Ow} values to predict bioconcentration by certain aquatic organisms.

5.2 Organic Pollutants in Terrestrial Ecosystems

The routes by which pollutants reach the land surface and terrestrial animals and plants were described in Chapter 2. Pollutants may remain on the land surface, enter terrestrial food chains, or be transferred to air or water. Contamination of the land surface occurs where there is dumping e.g. at landfill sites or in the vicinity of industrial premises, and in soils. Agricultural soils are of particular interest and concern because they receive treatments with pesticides which, by definition, have high toxicity to certain types of organism. Once pesticides enter soils, questions arise about residues in crops, contamination of drinking water, and possible effects on soil organisms and soil fertility. The fate of organic pollutants in soil will be discussed before dealing with their movement through terrestrial food chains.

5.2.1 *Fate in Soils*

Pesticides are the most important pollutants in agricultural soils, and these may reach the soil directly, or by transfer of residues from plants to which they have been applied. Pesticides are applied as sprays, granules or dusts. Some pesticides are sufficiently soluble to be fully dissolved in spray water, but in most cases they are formulated as emulsifiable concentrates (dissolved in an oily liquid) or wettable powders (fine particles mixed with inert diluent) because of their low water solubility. In this latter case, droplets (emulsifiable concentrate) or particles (wettable powder) are suspended in water. The availability of the pesticide is dependent on formulation; on rates of release from particles, droplets or granular formulations. Apart from pesticides, soils are sometimes contaminated by hydrocarbons, PCBs and other industrial chemicals. Such compounds are usually far less toxic than pesticides, and pollution problems associated with them are more localized. The fate of pesticides and other organic pollutants in soils will now be discussed, before considering their transfer along food chains.

Soils are complex associations between living organisms, mineral particles and organic matter. The clay fraction of the minerals, and the so called humus of the organic matter are colloids (diameter < 0.002 mm), which on account of their small size, present a large surface area per unit volume. When organic compounds enter soils, they become distributed between soil water, soil air and the available surfaces of soil minerals and organic matter. Where they are in the form of – or associated with – particles or droplets, some time will elapse before the individual molecules are distributed between these compartments of the soil. This will be the case, for example, where pesticides formulated as wettable powders are applied as sprays.

The distribution of organic compounds in soil is dependent upon their physical properties – especially solubility (e.g. K_{Ow} values), vapour pressure and chemical stability (Figure 5.12). Taking chemical stability first, organic pollutants are broken down by hydrolysis, oxidation, isomerization and (if on the surface) the action of light (photochemical breakdown). Usually breakdown leads to a loss of toxicity, but occasionally the products are themselves highly toxic (e.g. the isomerization of the organophosphorus compound malathion to isomalathion). Polar compounds (hydrophilic compounds of low K_{Ow}) tend to dissolve in water, and are adsorbed to soil colloids only to a limited degree. (Sometimes organic compounds which exist as ions provide an exception to this rule, because they become strongly associated with sites on soil colloids which bear an opposite charge. Thus the herbicide paraquat exists as a cation, and binds strongly to negatively charged sites of clay minerals.) Conversely, compounds of low water solubility (high K_{Ow}) tend to become strongly adsorbed to the surfaces of clay and soil organic matter, but exist only at very low concentrations in soil water. Compounds of high vapour pressure tend to volatilize into the soil air or directly from the soil surface into the atmosphere. If they enter soil air, they may be retained within the soil for some time but will eventually pass into the atmosphere.

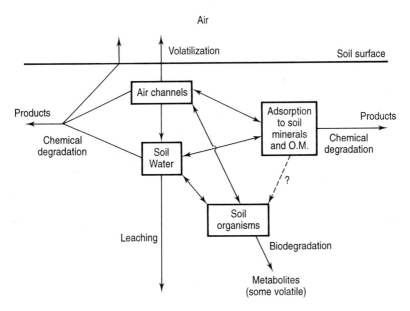

Figure 5.12 The fate of pollutants in soil.

The binding of organic molecules to soil colloids restricts their movement in the soil, and their availability to soil organisms. Thus, compounds of high K_{Ow} (i.e. lipophilic compounds) applied to soil show little tendency to be leached down through the soil profile by water. Some soil-acting herbicides (e.g. simazine) exhibit depth-selection in their herbicidal action, because of this phenomenon. They are only toxic to surface rooting weeds, because they are not carried far enough down the profile to be taken up by deeper rooting ones. Limitation of availability to soil organisms restricts the rate of biotransformation, and the toxicity of lipophilic compounds. Thus chemically stable lipophilic compounds often have long half lives in soil, because they are tightly bound to clay and/or organic matter, and are metabolized very slowly.

By contrast, hydrophilic compounds (e.g. soluble herbicides such as 2,4D and MCPA), which are not strongly bound to soil clay or organic matter, move more freely in soil and are readily available to soil organisms. They tend to be carried down the soil profile by water. Also they tend not to be very persistent, because soil organisms metabolize them relatively rapidly.

Active forms of organochlorine insecticides such as p,p'DDT and dieldrin provide examples of lipophilic compounds which are only slowly metabolized. Even when freely available, they are metabolized only slowly because they are poor substrates for detoxifying enzymes. The loss of compounds such as these from soils is biphasic (Figure 5.13). Immediately after application there is a period of relatively rapid loss when newly applied insecticide, which is present as particles or dissolved in an oily solvent, is volatilized or simply blown away as dust. During this period molecules of the insecticide become adsorbed to soil colloids, and the period of initial rapid loss is succeeded by a period of slow exponential loss. This is because the adsorbed insecticide only slowly becomes available for loss by evaporation or metabolism. In the period of slow loss the half lives for these compounds can run into years – in some cases even tens of years (see Table 5.5). The half lives depend on the compound, the type of soil and the climate. Compounds with higher vapour pressure tend to be lost more rapidly than compounds of lower vapour pressure. Compounds which are biotransformed rapidly tend to be lost more rapidly than compounds which resist biotransformation. Persistence tends to be greatest in

Table 5.5 Persistence of organochlorine insecticides in soils[a,b]

Compound	Time for 50% loss in years	Time for 95% loss in years
Dieldrin	0.5–4	4–30
p,p'DDT	2.5–5	5–25
Lindane (γBHC)	(approx. 1.5)	3–10

[a] After Walker (1975).

[b] Since the rate of loss from soils falls into different phases, true half lives cannot be determined. However, estimates have been made of the time taken for a particular percentage of an applied dose to disappear.

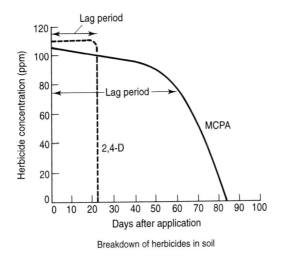

Breakdown of herbicides in soil

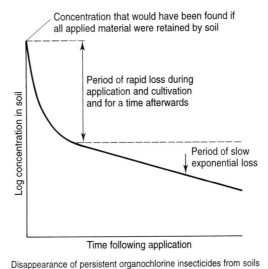

Disappearance of persistent organochlorine insecticides from soils

Figure 5.13 Loss of pesticides from soil. From Walker (1975).

heavy soils which are high in clay and/or organic matter. Persistence is also favoured by low temperatures. Under tropical conditions, rates of loss due to volatilization, chemical breakdown and biotransformation tend to be faster than under cooler, more temperate conditions.

Hydrophilic compounds such as the herbicides 2,4D and MCPA, and the insecticide carbofuran show a fundamentally different pattern of loss from that just described (Figure 5.13). Immediately after application to soil, the rate of

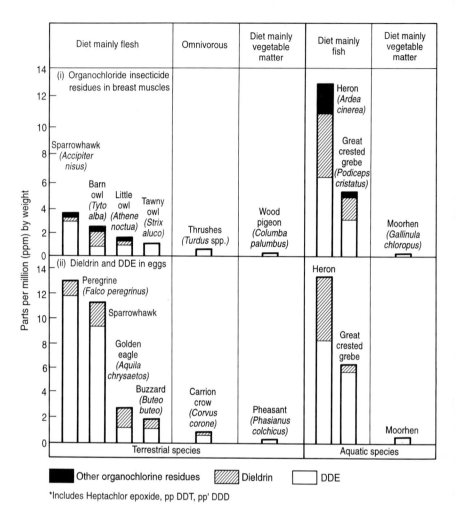

Figure 5.14 Organochlorine insecticides in British birds. From Walker (1975).

loss is relatively slow. However, after a period of time (lag phase) the rate becomes much more rapid and the compound is quickly lost. Typically the chemical disappears within a period of days or weeks. If more of the compound is immediately added to the soil, then this also quickly disappears. The soil has become 'enriched'. Strains of microorganisms have developed in the soil, which can metabolize the organic compound. If no more compound is added to the soil, this 'enrichment' will be lost, and the soil will revert to its original state. This enrichment phenomenon is sometimes a problem in agriculture; certain pesticides lose their effectiveness if they are used too often (carbofuran is a case in point).

The fate of organic compounds in soil – their movement and their decay curves – have been predicted with some success using models which incorporate parameters such as K_{Ow} and vapour pressure.

5.2.2   *Transfer along Terrestrial Food Chains*

Organisms in terrestrial ecosystems may take up pollutants from their food and ingested water, or directly from air, water or solid surfaces with which they come into contact. Uptake from ambient water is not such an important route as it is for aquatic organisms, although it should be borne in mind that soil organisms (e.g. earthworms) may acquire pollutants in this way. Uptake from ingested food or water is a very important route of uptake – often the major route for terrestrial vertebrates. Here, passage of organic pollutants, or their stable biotransformation products, along food chains is a matter of great importance. Where compounds have long biological half lives, their passage along a food chain may lead to biomagnification, at some or all of the stages. Typically, the highest concentrations of pollutant will be found in predators of the higher trophic levels of the food pyramid. Also, because of the mobility of some animals (especially birds), they may be transported to areas far removed from the point where they were originally released e.g. between Africa and Europe. For a discussion of the problem, see Balk and Koeman (1984).

The existence of relatively high levels of persistent lipophilic pollutants in terrestrial predators has often been reported. Examples include organochlorine insecticides (Figure 5.14), PCBs, and methyl mercury. There is a shortage of reliable data on the concentrations of persistent pollutants in organisms of different trophic levels of the same terrestrial ecosystem at the same time.

Levels of persistent organochlorine insecticides were determined in organisms from different trophic levels of terrestrial ecosystems in Britain during the 1960s. Like data from the marine environment (Figure 5.15 and section 5.3), these point to a general upward trend in residues of compounds like *p,p′*DDE (metabolite of *p,p′*DDT) and dieldrin with movement along the food chain. However, there is a danger of oversimplifying the complex situation that existed in the field at a time when there were very large temporal and spatial variations in exposure to these compounds. For example, grain-eating birds acquired lethal doses of dieldrin in some areas (e.g. agricultural areas of Eastern England), where they could feed on grain that had been dressed with it (Turtle *et al.*, 1963; Moore and Walker, 1964); in nearby areas, where the chemical was not used, the exposure of the same species was practically zero. Also, there was not necessarily biomagnification at each step in the food chain, even for highly persistent compounds. Dressed seed might have a dieldrin residue of 100 μg g^{-1} – but a grain-eating bird or mammal could not normally build up such a concentration in its tissues because lethal dieldrin poisoning would normally occur at levels well below this. Further, species such

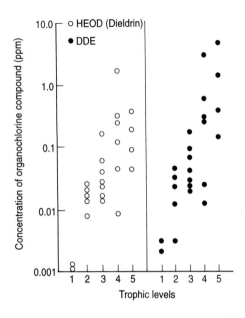

Trophic levels in Farne Island ecosystem:
1. Serrated wrack, Oar weed. 2. Sea urchin, Mussel, Limpet. 3. Lobster, Shore crab, Herring, Sand eel. 4. Cod, Whiting, Shag, Eider duck, Herring gull. 5. Cormorant, Gannet, Grey seal.

Figure 5.15 Organochlorine insecticides in the Farne Island ecosystem. From Walker (1975).

as the sparrowhawk and peregrine tend to feed selectively. Thus, in the situation under consideration, they may have tended to select prey with the highest residue concentrations in their tissues – i.e. individuals showing sublethal symptoms of poisoning. They may have been exposed, not to average concentrations of pollutant in the tissue of the prey species, but to the highest concentrations in surviving members of that species.

One important issue is the tendency of predators to bio-accumulate persistent pollutants with long biological half lives present in prey, when exposed to them over long periods. Available data suggest that predatory birds exposed to compounds like dieldrin or *p,p'*DDE can achieve bio-accumulation factors of 5–15 fold in relation to their prey, if a steady state is reached. As mentioned earlier, predatory birds are thought to be particularly efficient bioaccumulators of lipophilic xenobiotics because they have poorly developed oxidative detoxication systems.

In summary, certain persistent lipophilic pollutants have been shown to be transferred along terrestrial food chains, sometimes reaching their highest concentrations in predators of the highest trophic levels. There has been a major problem with certain pesticides used as seed dressings (e.g. aldrin, dieldrin, methyl mercury) where the concentration at the first step of the food chain is already high and grain-eating species can quickly acquire lethal doses if they feed on dressed grain. The persistence of most of these compounds is due to a combination of high lipophilicity (high K_{Ow}) and resistance to metabolic detoxication (especially in species such as specialized predators which are deficient in detoxifying enzymes anyway). Lipophilic compounds that are readily metabolised tend not to be persistent in terrestrial animals. If they have toxic effects, these seem to be 'acute' in nature – of short duration and limited to the species immediately exposed to them, and to the area in which they are released.

5.3 Organic Pollutants in Aquatic Ecosystems

As in the case of soils, the fate of organic pollutants entering aquatic ecosystems depends on their physical properties, especially lipophilicity, vapour pressure and chemical stability. Compounds which lack stability – e.g. compounds that tend to be hydrolyzed – present few problems as aquatic pollutants unless their transformation products happen to be toxic. Also, volatile compounds tend not to persist in aquatic ecosystems. Polarity is once again important in determining distribution and persistence. Hydrophilic compounds tend to be dissolved in and distributed throughout surface water. Lipophilic compounds on the other hand, tend to become associated with particulate matter, notably of sediments. Sometimes they may also exist on the surface of water, e.g. dissolved in surface oil films.

5.3.1 *Pollutants in Sediments*

Like soils, the sediments of rivers, lakes and seas are associations between organic and inorganic particles and living organisms. Organic pollutants may be adsorbed to the particles of sediments, and this limits their mobility and their availability to bottom-dwelling organisms. It is often uncertain to what extent pollutants can be taken up directly by animals from the adsorbed state, or whether they need first to move into the aqueous medium before they become available. The extent to which an adsorbed pollutant can be taken up directly will depend upon the nature of the chemical, the nature of the surfaces to which it is bound, the strength of the binding, the species which is taking it up and – in some cases at least – the temperature, pH and oxygen content of the ambient water.

The oxygen content of water can be important in determining the nature and the rate of both chemical and biochemical transformations. As the oxygen content declines, so there will be a tendency for oxidative transformations to be replaced by reductive ones (section 5.1.5). Sediments can differ widely in regard to their oxygen content. At the bottom of deep seas, conditions are anaerobic, whereas in shallow, fast-moving streams oxygen levels are relatively high. Sediments of intertidal zones along the sea shore experience fluctuating oxygen levels, in accordance with tidal movements.

5.3.2 Transfer along Aquatic Food Chains

The same general considerations apply as in the case of terrestrial food chains (section 5.2.2) except that exchange diffusion between organisms and ambient water is a complicating factor in aquatic food chains. This makes more diffi-cult the interpretation of data such as those shown in Figure 5.15. The residue levels shown for the persistent organochlorine compounds dieldrin and p,p'DDE (metabolite of p,p'DDT) were measured in organisms sampled from different trophic levels of the marine ecosystem around the Farne Islands, Northumberland, Great Britain, during the period 1962–1964 (Robinson *et al.*, 1967). This shows a strong relationship between the log concentrations of the residues, and the trophic levels of the organisms in which they were deter-mined. The lowest levels are in the plants (brown algae) in trophic level 1, the highest in the vertebrate predators of trophic levels 4 and 5. It is interesting that p,p'DDE is subject to a steeper gradient than dieldrin. p,p'DDE is metab-olized more slowly and has a longer biological half life than dieldrin in the animals that have been studied. It appears as if biomagnification occurred at every stage of the food chain. However, aquatic invertebrates and fish in trophic levels 2–4 obtained an unknown proportion of their residue burden directly from water and/or sediment and not from food. Also with predators (as mentioned previously) there may be selective predation. The prey that they eat may contain higher organochlorine residues than the average values shown in Figure 5.15. These points aside, the predators appear to be achieving a substantial biomagnification (i.e. a BF considerably greater than 1) assuming that most of the pollutant burden comes from food. The apparent BF calcu-lated from the data given in Figure 5.15 lies between 50 and 60 for the fish-eating shag (*Phalacrocorax aristotelis*) (Walker, 1990a). The variation in residue levels in the principal prey species, the sand eel (*Ammodytes lanceolatus*), was not large. Even if they were feeding on the most contami-nated individuals, a BF of several fold is still suggested.

The fish-eating birds shown to bio-accumulate substantial levels of persist-ent organochlorine compounds in this and other studies, have low activities of the monooxygenase system, which is mainly responsible for their detoxication (Walker, 1990a; Ronis and Walker, 1989; Figure 5.6). The species in question include the shag, cormorant (*Phalacrocorax carbo*), guillemot (*Uria aalge*) and razorbill (*Alca torda*). In general, a deficiency in their detoxication system favours bio-accumulation.

There is a general concern about the build up of relatively high levels of persistent pollutants in predators at the top of aquatic food chains (for a recent review of the question see Walker and Livingstone (1992)). Apart from the organochlorine insecticides just mentioned, persistent polychlorinated biphenyls (PBBs) and polychlorinated dibenzodioxins (PCDDs) have also given rise to concern. In addition to sea birds, marine mammals such as seals and cetaceans (porpoises, dolphins and whales) have been shown to bio-accumulate these compounds – and to pass them to their offspring via milk. Like sea birds, predatory marine mammals have poorly developed mono-oxygenase detoxication systems. The above trend has also been observed in freshwater ecosystems where relatively high levels of persistent organochlorine compounds have been found in predatory birds (e.g. heron (*Ardea cinerea*)) and mammals (e.g. otter (*Lutra lutra*)) in Western Europe, and in predatory birds (e.g. double-crested cormorant (*Phalacrocorax auritus*)) of the Great Lakes of Canada and the USA.

Lipophilic pollutants which have relatively short half lives (e.g. polycyclic aromatic hydrocarbons, PAHs) do not show the same tendency to pass along food chains and to be biomagnified. Some invertebrates of the lower trophic levels (e.g. the mussel *Mytilus edulis*) bioconcentrate and/or bio-accumulate them – because they have little ability to metabolize them. On the other hand, fish, birds and mammals metabolize them rapidly by monooxygenase attack, so there is little tendency for them to reach the higher trophic levels let alone to be bio-accumulated there. Although these compounds do not raise the problem of biomagnification, it must be emphasized that some PAHs and other readily degradable compounds, are subject to metabolic activation (section 5.1.5).

Further Reading

HODGSON, E. and LEVI, P. (1993) *Introduction to Biochemical Toxicology*, 2nd Edition. A multi-author text covering the main aspects of biochemical toxicology in reasonable depth. Covers toxicokinetics of xenobiotics and mode of action of 'poisons'.

TIMBRELL, J. A. (1992) *Principles of Biochemical Toxicology*, 2nd Edition. A single-author text giving a very clear and readable account of the basic principles of the subject.

WALKER, C. H. and LIVINGSTONE, D. R. (1992) *Persistent Pollutants in Marine Ecosystems* Gives a detailed account of the fate and levels of persistent organic compounds in marine food chains.

MORIARTY, F. (1975) *Organochlorine Insecticides: Persistent Organic Pollutants* Probably the best researched of all the persistent organic pollutants. Much information on their fate and environmental behaviour.

ENVIRONMENTAL HEALTH CRITERIA A range of monographs on particular pollutants published by the World Health Organization (over 160 titles). Some of them give detailed reviews of the fate of particular pollutants. Selected titles are in the references at the end of this book.

PART TWO

Effects of pollutants on individual organisms

6

Toxicity Testing

The first five chapters have been concerned with the fate of chemicals in the environment, and within living organisms. This chapter begins the second part of the text, where the emphasis shifts to questions about the effects that chemicals have upon individual organisms. Ecotoxicology is concerned with the harmful effects of chemicals. This chapter will address the questions: what constitutes harm, and how is toxicity measured? It is important to do this before moving on to the more complex issues of Part Three where the relationship between toxic effects upon individuals and consequent adverse effects at the level of population becomes a crucial question.

6.1 General Principles

Of central importance in both toxicology and ecotoxicology is the relationship between the *quantity* of chemical to which an organism is exposed, and the nature and degree of consequent harmful (toxic) effects. Dose–response relationships provide the basis for assessment of *hazards* and *risks* presented by environmental chemicals (more later). This simple basic concept immediately raises questions about the definition of poisons, since everything depends on dose. Paracelsus (1493–1541) recognised the dilemma and stated 'All substances are poisons; there is none that is not a poison. The right dose differentiates a poison and a remedy' – or, in other words, 'the dose maketh the poison'. Thus, no chemical is poisonous if the dose is low enough, while all chemicals are poisonous if the dose is high enough (even apparently harmless substances like sugar and salt can be toxic to animals at high doses). It is advisable to remember this principle when reading sensational popular articles on the subject, where reference is often made to 'poisonous' or 'toxic' substances in the environment, without pointing out that the levels are far too small to have any toxic effect.

There are many different ways in which toxicity can be measured. Most commonly the measure ('end point') is death, although there is a growing interest in the use of more sophisticated indices. The desire to minimize lethal toxicity testing with vertebrate animals is a significant driving force here. Biochemical, physiological, reproductive and behavioural effects can all provide measures of toxicity. Many toxicity tests provide an estimate of the dose (or the concentration in food, air or waters) which will cause a toxic response at the 50% level, e.g. median lethal dose, the dose that will kill 50% of a population. It is also possible – and this approach is gaining in popularity – to establish the highest concentration or dose that will *not* cause an effect.

Several terms used in relation to toxicity testing require definition. First, in lethal toxicity testing, LD_{50} represents the medium lethal dose, while LC_{50} represents the medium lethal concentration. In toxicity tests which determine these values, it is also possible to determine the highest doses or concentrations which cause no toxicity – the No Observed Effect Dose (NOED) and No Observed Effect Concentration (NOEC) respectively. These values can only be determined in situations where a higher dose or concentration has produced an effect in the same toxicity test. These points are illustrated in Figure 6.1, which refers to the determination of an LC_{50} after 96 h exposure. If a test is carried out where the end point is an adverse response other than death, then an EC_{50} or ED_{50} is determined. Here the concentration or dose producing the effect in 50% of the population is determined. As with lethal toxicity testing, it is possible to determine NOEC or NOED, following this approach.

Apart from toxicity tests involving the use of live animals which are the principal subject of this chapter, there are other ways of evaluating the toxic

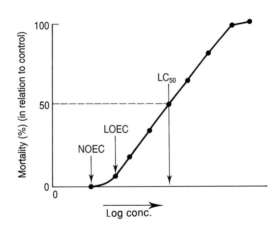

Figure 6.1 Toxicity after 96 h exposure in an aquatic toxicity test. It should be noted that NOEC can be determined only where LOEC is known – otherwise there would be no indication of a concentration that can be toxic. NOEC = no observed effect concentration; LOEC = lowest observed effect concentration; LC_{50} = median lethal concentration at 96 h.

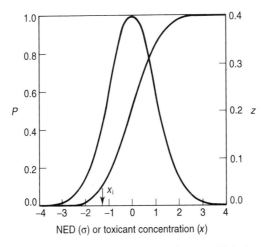

Figure 6.2 Plot of standardized normal frequency distribution with its integral, the cumulative normal frequency function.

6.2 Toxicity Testing with Terrestrial Organisms

6.2.1 Introduction

It is sometimes stated that toxicity testing with terrestrial animals is simpler than with aquatic animals as only one route of exposure via the gut has to be considered. Although this is the case with widely used tests with rats and mice, this is certainly not the case with more recently developed tests with invertebrates where exposure via the external medium (air, soil) is important also. In this section, ecotoxicological tests with invertebrates, plants and birds will be described.

6.2.2 Invertebrates

Two ecotoxicological tests using earthworms and bees have been developed to test the effects of chemicals on terrestrial invertebrates. Several others are at various stages of formulation at the time of writing. The methods of testing on four types of organism will be covered here, a widely-used standard test with earthworms, a test currently undergoing international validation with spring-tails, a test in the early stages of development with woodlice, and a standard test using bees. The earthworm test mainly measures effects of chemicals that pass across the body surface. The springtail test measures effects on repro-duction, again mainly through contact poisoning of the adults, their eggs, or juveniles that hatch from the eggs. The isopod test measures the effects of chemicals on feeding rates, mainly through feeding repellence. The bee test is designed to assess the effects of chemicals on 'beneficial insects'.

113

6.2.2.1 Earthworms

The OECD earthworm test is the best-established method and has reached the stage of being a legislative requirement in some countries before new chemicals can be released into the environment. Several approaches have been adopted (described in detail in Greig-Smith *et al.*, 1992). The most widely used species is *Eisenia fetida* (Figure 6.3A) which is easy to culture in the laboratory and has a reproductive cycle of about 6 weeks at 20°C. However, this is a species native to the southern Mediterranean which survives only in compost or manure heaps, or heated glasshouses during the winter in northern latitudes. Thus, care should be taken in extrapolating the results of tests with *Eisenia* to other species of earthworm.

The basic tests uses an artificial soil made up from 70% sand, 20% kaolin clay and 10% Sphagnum peat (by weight). Water is added to comprise 35% (by weight) of the final soil and the pH is adjusted to 6.3 with calcium carbonate. Ten adult worms are added to each of four replicates of a control, and an ascending series of concentrations of the chemicals under test. The worms are left for 14 days after which the number of survivors is counted. An LC_{50} for survival can then be calculated. Some workers have run their experiments for a longer period to allow the worms to reproduce (e.g. Spurgeon *et al.*, 1994). The reproductive rate is easy to assess by counting the number of cocoons produced (Figure 6.3B). Effects on reproduction are detected at lower concentrations of metals in soils than effects on growth and survival and it is thought that the former approach is more ecologically relevant in trying to assess potential effects in the field (Figure 6.4). Interestingly, there is also evidence that the artificial soil may be deficient in copper (Figure 6.4C) although further work is required to substantiate this suggestion.

One problem with the standard OECD method is that the worms have no food during the experiment and tend to lose weight. More recent studies have included a small pellet of cow or horse manure on the surface of the artificial soil which allow the worms to put on weight. Fed worms also have a higher reproductive rate than starved specimens. The bioconcentration factor (concentrations of the chemical in worms/concentration of the chemical in the soil) can also be determined from such experiments and can be used to predict the exposure of earthworm predators to potential toxins in the field.

Where the results of toxicity tests using laboratory soils have been compared with those in 'natural' soils brought in from the field, it is clear that chemicals are invariably much more 'bioavailable' to worms in the artificial medium. Indeed for metals such as zinc, worms are affected at concentrations five to ten times lower in artificial soils than in field soils (Spurgeon *et al.*, 1994). The artificial soil has a lower organic matter content than many field soils, and the clay used has a relatively low cation exchange capacity. In terms of 'environmental protection' therefore, the test includes an inbuilt 'safety factor' which should be born in mind when results of laboratory tests are being extrapolated to the field.

Figure 6.3 (A) Group of earthworms (*Eisenia fetida*) in standard OECD soil. Adult worms are approximately 8 cm in length. (B) Juvenile *Eisenia fetida* emerging from a cocoon of approximately 2 mm in diameter. Photographs copyright of Steve Hopkin.

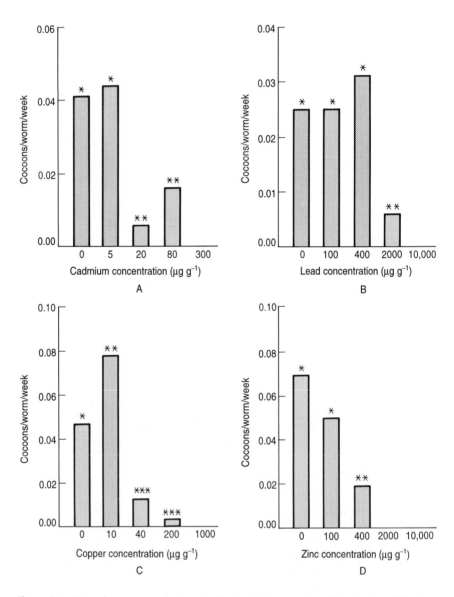

Figure 6.4 Rate of cocoon production by *Eisenia fetida* exposed to (A) cadmium, (B) lead, (C) copper, and (D) zinc ($\mu g\ g^{-1}$ dry weight). Bars with the same number of asterisks (*) were not significantly different at $P < 0.05$. Reproduced from Spurgeon et al. (1994) with permission of Elsevier Science.

6.2.2.2 *Springtails*

Springtails (Insecta: Collembola) are among the most abundant of all soil invertebrates. Although the springtail test has not reached such an advanced legislative stage as that for earthworms, many laboratories in the world use this species to assess the effects of chemicals on non-target soil arthropods.

The most widely used species is a parthenogenetic strain of *Folsomia candida* (Figure 6.5) which has a reproductive cycle of three to four weeks at 20°C. A similar experimental design to that used in the earthworm test is constructed with four replicates each of a control, and an ascending series of concentrations of the test chemical. The chemical is mixed in the same formulation of artificial soil as described for the earthworm test above. Ten adult springtails of the same age are placed in a small container about the same size as a plastic vending machine cup (about 10 cm in height, 6 cm in diameter) together with a small amount of dried yeast for food. They are left for a minimum of 3 weeks, then the soil is flooded. All the springtails, including any offspring produced during the experiment, float to the surface of the water. Each container is photographed and the developed film is projected onto a large screen on which the number of progeny are counted (Figure 6.6). The concentration which causes a 50% (or other percentage) reduction in reproduction compared to controls can then be calculated. If a larger number of replicates is prepared, then a proportion of the containers can be flooded at intervals to produce a timed response (Figure 6.7).

Figure 6.5 Adults and juveniles of *Folsomia candida*, a parthenogenetic springtail. The largest specimen is 2 mm in length. Photographs copyright of Steve Hopkin.

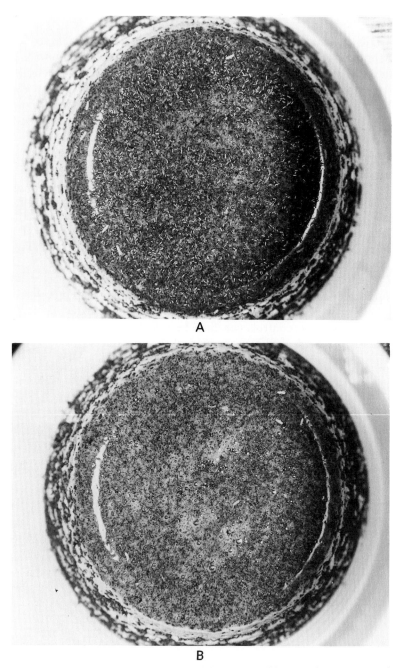

Figure 6.6 View from above of *Folsomia candida* maintained for 4 weeks in OECD artificial soil (A) or artificial soil containing 10 000 μg Pb g^{-1} dry weight (B). The containers of 8 cm in diameter were flooded with water immediately before the photographs were taken. Note the lack of juveniles in (B). Photographs copyright of Steve Hopkin.

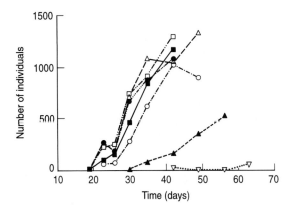

Figure 6.7 Total numbers of *Folsomia candida* individuals at different levels of cadmium in artificial soil. Blank (□), 34.8 (●), 71.3 (△), 148 (■), 326 (○), 707 (▲) and 1491 (▽) µg Cd g^{-1} dry weight. Reproduced from Crommentuijn *et al.* (1993) with permission of Academic Press.

6.2.2.3 *Woodlice*

Tests with woodlice (Crustacea: Isopoda, Oniscidea) are still at an early stage of development. One problem with species such as *Oniscus asellus* (Figure 6.8B) and *Porcellio scaber* is their low growth rate and long reproductive cycle relative to *Eisenia fetida* and *Folsomia candida*. Even at 25°C, it still takes a minimum of 6 months for newly hatched juvenile woodlice to reach reproductive age. However, woodlice are potentially important test animals as they are common and widespread and perform an important role in physically breaking down leaf material into smaller particles which are more easily decomposed by microbes.

Woodlice are extremely resilient to starvation and can survive for many weeks without food. However, one parameter that can be measured with relative ease is the feeding rate. Woodlice eat leaf litter and convert this to faecal pellets with very consistent shape and weight (Figure 6.8B). Thus, by counting the number of faecal pellets produced during specified time intervals, it is possible to determine whether the presence of a chemical on the leaves is deterring the woodlice from feeding. Reduced feeding rate will have knock on effects such as slower growth rate and lengthened reproductive period. However, more experiments need to be conducted before woodlice can be considered eligible for 'standard toxicity test' status.

6.2.2.4 *Beneficial Insects*

Following concern as to the effects of pesticides on non-target insects, a number of laboratories test new chemicals on honey bees (Gough *et al.*, 1994).

119

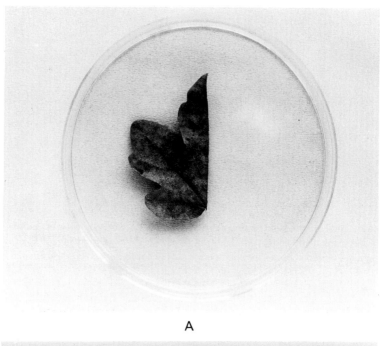

A

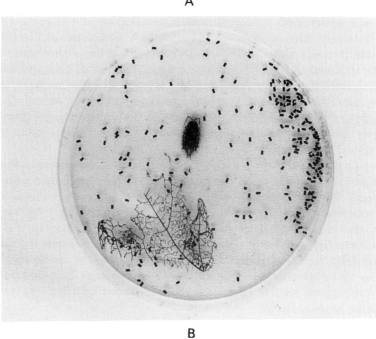

B

Figure 6.8 (A) Half a field maple leaf (*Acer campestre*) in a petri dish of 9 cm in diameter before addition of a woodlouse. (B) Three weeks later, the leaf has been reduced to faecal pellets by the feeding activity of *Oniscus asellus*. Photographs copyright of Steve Hopkin.

The standard method is to collect workers from a hive, anaesthetize them with carbon dioxide, then maintain the bees in cylinders of wire mesh (10 in each container) with three replicates of each test concentration plus controls. Each container is supplied with a sugar solution in a feeding bottle containing a known concentration of the chemical being tested. The cages are kept at constant temperature (typically 25°C) and checked at 1, 2, 4, 24 and 48 h for mortality of bees. In an additional test, contact toxicity is measured by applying the chemical directly to the thorax of bees.

A field-based method can also be used but this requires considerably more space. A bee hive is maintained in a large enclosed tunnel of polythene or netting and a pollen and/or nectar-bearing crop is sprayed with the test chemical. Bees which return to the hive but then die inside during the night are ejected in the morning by healthy workers. The level of mortality can thus be simply determined by counting the number of dead bees in a tray placed under the hive entrance.

6.2.3 *Vertebrates*

The toxicity of chemicals to mammals, birds and other vertebrates has commonly been measured as a median lethal dose (LD_{50}). In routine toxicity testing, a single dose is given orally to obtain an estimate of acute oral LD_{50}. To do this, groups of animals are given doses of the test chemical over a range of values which centre on a rough estimate of LD_{50} obtained in a preliminary range test. The percentage of animals which die in each group over a fixed period following dosing is then plotted against the log of the dose (mg kg^{-1}). To obtain a straight line relationship between dose and mortality, it is necessary to transform percentage kills into normal equivalent deviates (NEDs) or probit values (probit analysis) (Figure 6.9).

Values of LD_{50} are sometimes obtained using other methods of dosing. Chemicals may be injected into the blood, into the muscle, or into the peritoneal cavity, or they may be applied directly to the skin (dermal LD_{50}). Sometimes they are continuously administered in the food or water over a fixed period of time. In this case, toxicity is expressed as a 'median' lethal concentration in food or water over a stipulated period, e.g. a 5-day oral LC_{50}.

During the course of this type of toxicity testing, it is possible to establish values for no observed effect dose (NOED), i.e. the highest dose given that produced no lethal effect. Nowadays there is a movement away from the use of the LD_{50} tests on ethical grounds. The British Toxicology Society has proposed an alternative procedure which requires the use of far fewer animals. This would merely classify compounds as 'harmful', 'toxic', or 'very toxic' (for further details see Timbrell, 1995).

6.2.4 *Plants*

A wide variety of methods have been developed for assessing the toxicity of chemicals to plants. In this section, tests to assess the effects of metals will be

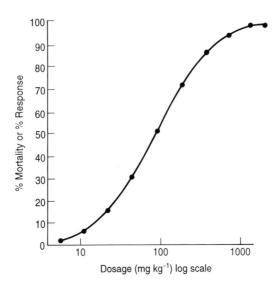

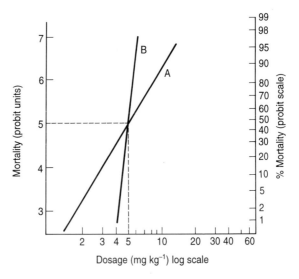

Figure 6.9 Determination of LD_{50}. For details see text.

covered in most detail as this is where most work has been concentrated. However the basic principles of these tests are applicable to other chemicals, and in tests to assess the effects of acid rain and gaseous pollutants.

Plants which naturally contain very high concentrations of specific metals are known as *metallophytes* (Baker and Proctor, 1990). Some metallophytes are able to grow naturally on metal-contaminated soils. Other non-metallophytes have evolved genetically distinct strains which are resistant to metals and are able to grow much better in contaminated situations than non-resistant plants of the same species (Schat and Bookum, 1992). Resistant plants may also grow more successfully in nutrient poor conditions which often exist in acidic mine waste (Ernst *et al.*, 1992) (see section 13.6). There has been considerable interest in using tolerant plants to revegetate old mine sites (Plenderleith and Bell, 1990).

A number of techniques have been developed for quantifying metal tolerance in plants (Baker and Walker, 1989). The degree of tolerance is measured by comparing the response to tolerant plants (from contaminated sites) and expressing this as a percentage of the performance of non-tolerant plants (collected from 'clean' areas). The principal parameters measured are seedling survival, biomass, shoot growth and root growth.

Seedling survival is the number of plants surviving from seed after a specified time period. It is a better measure than straightforward germination. Many non-resistant plants will successfully germinate in metal-contaminated soil but subsequently they fail to grow and remain characteristically stunted.

Rates of dry matter production and the biomass yield of resistant plants are generally found to be lower than their non-resistant counterparts when grown in uncontaminated soil. This reduction is believed to be due to the energy expenditure in metal-tolerance mechanisms, such as compartmentalization of metals in intracellular 'compartments' (Vasquez *et al.*, 1994) (see section 13.6). When grown in contaminated soil or nutrient solutions, the growth of resistant plants exceeds that of non-resistant ones. Other ways of measuring this difference are shoot and root length. Shoot length can be compared in plants grown in soil but measuring differences in root length is much easier if the plants are reared in nutrient solutions in clear containers. Regular monitoring of the root length can be conducted without disturbing the plants. This approach has been used in numerous demonstrations of resistance as it allows the composition of the liquid medium to be closely controlled (Table 6.2).

Table 6.2 Percentage tolerance of three species of grass collected from Hallen Wood and Midger Wood, an uncontaminated woodland[a]

	Hallen plants	Midger plants
Dactylus glomerata	105.9	57.4
Holcus lanatus	113.8	28.1
Deschampsia	82.2	37.5

[a] The tolerance was calculated from the mean length of roots in 2 ppm Cd/mean length of roots with no Cd. The measurements were made after 14 days in full strength Hoaglands culture solution. From Martin and Bullock (1994).

Deleterious effects on plants of gaseous pollutants such as ozone can be detected by methods in addition to those reported above. Yellowing of leaves, *chlorosis*, is a characteristic feature of stress. The critical air concentrations can be determined by exposing replicates to different levels of the gas under test in separate chambers, although such experiments are expensive to conduct. The degree of chlorosis can be measured by counting the number of leaves affected, or by determining the area of leaf exhibiting the yellow coloration (Mehlhorn *et al.*, 1991). Different varieties of the same species may exhibit different degrees of chlorosis at the same chemical concentration and may be useful for biological monitoring of pollution (e.g. tobacco, see Chapter 11).

6.3 Toxicity Testing with Aquatic Organisms

The basic principles of aquatic toxicity testing are similar to those already described for terrestrial organisms. However, there are particular questions about the main routes of uptake which influence some aspects of the design of tests.

With aquatic organisms, direct uptake from water is a route of major importance (e.g. uptake across the gills of fish, or across the permeable skin of amphibia). Uptake can also occur from food during its passage through the alimentary system. Bottom-dwelling organisms are exposed to residues in sediment. The relative importance of these routes of uptake differs between organisms and between chemicals and depends on environmental conditions. In some cases all of these routes may operate in one organism at one time.

Much of the toxicity testing carried out with aquatic organisms (e.g. fish, *Daphnia*, *Gammarus pulex*) is concerned with direct absorption from water. Organisms are exposed to different concentrations of the chemical in water, in order to determine values for median lethal concentration. Although absorption is primarily direct from water, it is not possible to completely exclude contamination of food, and thus some uptake from this source.

One difficulty in aquatic toxicity testing is maintaining a constant concentration of chemical in water. Chemicals are lost from water due to (1) absorption and metabolism by the test organism, and (2) volatization, degradation and adsorption from water. Where the rate of loss is relatively low, tests may be performed using *static* or *semi-static* systems. With static systems, the water is not changed for the duration of the test. With semi-static systems, the water is replaced at regular intervals (usually every 24 h). A better, but more complex and expensive, method for renewing test solutions is provided by a *continuous-flow* (*flow-through*) system. With systems of this type, test solution is continuously renewed, thus ensuring a constant concentration of the test chemical, and preventing the build-up of contamination from faeces, algae, mucus etc. If organisms are exposed to a chemical for a sufficiently long time, steady state concentrations will be reached in the tissues (Chapter 5).

The toxic effect of a chemical depends upon the concentration present in the tissue(s) where the site of action is located (Chapters 5 and 7). This, in turn, depends upon the concentration of the chemical in water and the period of time over which exposure has occurred. Thus the median lethal concentration (LC_{50}) is related to the exposure period (Figure 6.10). With increasing exposure time the LC_{50} becomes less until the median lethal threshold concentration (threshold LC_{50}) is reached. At this point further increases in exposure period cause no change in mortality. It may reasonably be supposed that when the threshold LC_{50} is reached the system is in a steady state, i.e. the tissue concentration is no longer increasing with time.

In performing aquatic toxicity tests, preliminary ranging tests are necessary to obtain a rough estimate of toxicity. Here small groups of test organisms (typically two or three individuals per group) are exposed to a wide range of concentrations of the test chemical on a log scale. The results of the ranging test can be used to plan a full scale toxicity test, where larger numbers of test animals are exposed to a narrower range of concentrations, centring on the LC_{50} estimated from the ranging test. The percentage mortality in each of the test groups is recorded over various time intervals for the duration of the test. The data obtained can be used to determine the LC_{50} at different exposure times.

Initially, the data can be plotted in two ways:

1. At one fixed exposure concentration, a series of chosen survival times can be plotted against the percentage of the sample surviving at each of them (e.g. 25% of the sample survive for 4 days). This can then be repeated for

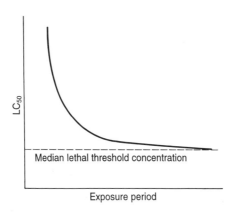

Figure 6.10   Relationship of LC_{50} to the exposure period.

each of the remaining concentrations to obtain values for the median survival periods (median periods of response).

2. At one fixed exposure period, the percentage mortality can be related to the exposure concentration. From this, the median lethal concentration can be calculated for each of the individual exposure periods (Figure 6.11).

From these data two further plots can be produced. Using data from 1, median survival time can be plotted against concentration of chemical. Using data from 2, the median lethal concentration can be related to exposure period, and from this LC_{50} values can be estimated for particular exposure periods e.g. LC_{50} at 96 h.

Up to this point the discussion has been restricted to lethal toxicity testing. It should be emphasized, however, that end points other than lethality may be employed. More generally median effective concentrations (EC_{50}) can be determined for a variety of non-lethal end points.

Since the toxic effect of a chemical is highly dependent upon the period of exposure, two important issues are: (i) over what period should a toxicity test

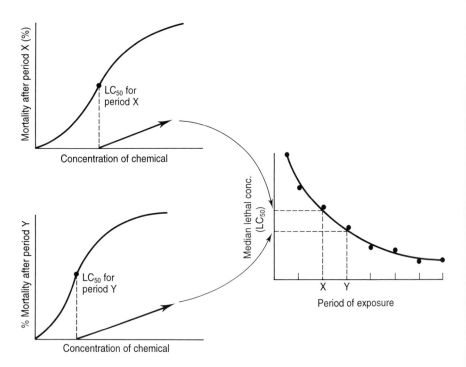

Figure 6.11 Determination of LC_{50}. For details see text.

be conducted? and (ii) what exposure periods should be chosen for the estimation of EC_{50} or LC_{50}? The answer to both of these questions depends on the organism being tested, and the purpose of the test. Tests with *Daphnia* are commonly of only 24–48 h duration in static systems. By contrast, fish toxicity tests are usually of longer duration. Simple screening tests commonly require relatively short exposure periods. Typical measures of toxicity are LC_{50} at 48 h for *Daphnia*, and LC_{50} at 96 h for fish. Such data give an indication of the toxicity of one chemical in relation to others (i.e. a ranking of toxicity). On the other hand, longer exposure periods may be required where toxicity data are needed for the evaluation of water quality. Here, exposure may be continued until a median threshold concentration has been established (Figure 6.9).

Some aquatic toxicity tests work to non-lethal end points. The *Daphnia* reproduction test is an example. Young female *Daphnia* are exposed to a chemical over a 3-week period in a static renewal system. Test results are based on the comparison of the number of offspring per surviving female to the reproductive output of controls. NOECs can be calculated from these results.

The essential feature of an algal toxicity test is to determine the effects of a chemical on a green algal population growing exponentially in a nutrient-enriched medium for 72 or 96 h (Lewis, 1990). Cell density is determined either using a direct measurement (microscopic counting) or one of several indirect techniques (spectrophotometric method or electronic particle counter). Controlled experimental factors include the initial cell density, test temperature, pH, light quality and intensity. Typically, an EC_{50} value and the NOEC are calculated based on growth inhibition. The EC_{50} value is the concentration of the test substance that causes a 50% reduction of the effect parameter.

The most frequently used test species is the freshwater green coccoid *Selenastrum capricornutum*, which is the only species recommended by the US Environmental Protection Agency for monitoring the toxicity of effluents. At least 10 other species have been used. However Lewis (1990) concluded that these are too few data supporting the environmental relevance of results using the standard method in their present format. The main problem is that interspecific variation in response to chemicals is significant and that field-validation of most laboratory-derived results is lacking (a problem common to most tests). The way forward is probably to conduct simultaneous multiple species tests to assess the relative toxicities of different chemicals to different species.

A number of factors influence the values obtained in aquatic toxicity testing, apart from the inherent properties of the chemical and of the test organism. A critical question is the availability of the chemical to the test organism. Availability may be limited because of adsorption to particulate matter or to the surface of the tank. The toxicity of chemicals associated with sediments is difficult to assess because it is uncertain how much of the total quantity is available to the test organism. Strongly adsorbed molecules may remain associated with sediment particules. Other factors which influence toxicity include temperature, pH and composition of water.

6.4   Risk Assessment

The toxicity data obtained by the testing procedures described here are eventually used to make assessments of hazard and risk (Calow, 1993, 1994; Walker *et al*, 1991b). For the purposes of this discussion the following definitions will be used:

hazard = the potential to cause harm
risk   = the probability that harm will be caused

Risk assessment depends on making a comparison between two things:

1. the toxicity of a compound expressed as a concentration (EC_{50}, LC_{50} or NOEC can be used);
2. the anticipated exposure of an organism to the same chemical, expressed in the same units as 1 (a concentration in water, food or soil to which the organism is exposed).

In hazard assessment, a toxicity test can give a plot which relates the frequency of a toxic effect (e.g. mortality) to the dose that is given (Figure 6.1). From this, an NOEC and an EC_{50} can be estimated. This can be compared to a putative 'high' environmental concentration to decide whether a hazard exists. A ranking of compounds according to their toxicity is important at this stage. If toxicity is very low, then a compound is not regarded as being hazardous.

In risk assessment, further calculations are carried out to obtain values for the predicted environmental concentration (PEC) and the predicted environmental no effect concentration (PNEC). The details of these calculations lie beyond the scope of the book. In the case of PEC, calculations are based on known rates of release and dilution factors in the environment. If, for example, a chemical is used on an industrial process, the level in the industrial effluent is measured or calculated. This figure is then divided by the dilution that occurs in receiving waters (e.g. river or lake) to obtain a value for PEC. The PNEC can be estimated by dividing LC_{50} or EC_{50} for the most sensitive species tested in the laboratory by an arbitrary safety factor (often 1000). This is to allow for the great uncertainty in extrapolating from laboratory toxicity data for one species to expected field toxicity to another species.

$$\frac{PEC}{PNEC} = \text{risk quotient}$$

If this value is well below 1, the risk is low. If it is 1 or above, there is a substantial risk.

Clearly, calculations such as these provide only rough estimates of risk. It is necessary to include a large safety factor to allow for uncertainties about environmental toxicity. Apart from the uncertainty over toxicity to particular organisms in the field, there is the further problem of estimating environmental exposure. In terrestrial ecosystems it is often very difficult to obtain a realistic value for PEC. If a mobile species (e.g. bird, mammal or insect) obtains a chemical mainly from its food – and the chemical is distributed very unevenly throughout the ecosystem – then it is not possible to estimate exposure with any degree of accuracy. A grain-eating bird may or may not consume grain treated with a pesticide; a mammal may or may not be in a field when it is sprayed with a pesticide.

Considerations such as these strengthen the case for the development of new strategies using biomarkers for the purposes of risk assessment. Biomarker assays can provide measures of exposure and sometimes of toxic effect under actual field conditions, e.g. in field trials with pesticides. This issue will be discussed further in Chapter 9.

6.5   Toxicity Testing in the Field

If the normal procedures of risk assessment raise uncertainties about the safety of a chemical, further testing may be necessary before a decision can finally be taken about permitted release into the environment. In the case of pesticides, field trials are sometimes required. In a full-scale field trial, a pesticide is likely to be applied at a dose, and under conditions which are most likely to produce toxic effects ('worst-case scenario'). A variety of measurements may be made, depending on the chemical, the method of application, the habitat, climate and agricultural system. These may include looking for dead animals and birds ('corpse-counting'), estimating population numbers and breeding success, measuring residues in soil, crop, and animals, and occasionally biomarker assays (e.g. cholinesterase inhibition, eggshell thinning in birds). Trials such as these are expensive, and are not undertaken lightly.

Box 6.1   Field study to assess environmental effects of ^{137}Cs on birds

One interesting recent study (Lowe, 1991) has demonstrated the importance of laboratory testing for field extrapolation if one is to prove or disprove a causative relationship between levels of an environmental pollutant and an ecological effect. This concerns an area in the vicinity of the Sellafield nuclear waste reprocessing complex in Northwest England.

Since 1983, concern had been expressed about the apparent decline in numbers of birds in the Ravenglass estuary in West Cumbria, particularly of the black-headed gull colony on the Drigg dunes. Suggestions had been made that the decline might be due to excessive radiation in the birds' food and their general environment. In Lowe's

(cont.)

(cont.) ─────────────────────────

(1991) study, 12 species of marine invertebrate from Ravenglass were analyzed. Most of them are known to be important food for birds but none of them showed excessive contamination with radionuclides.

Analysis of a sample of carcasses from the area showed that oystercatchers (*Haematopus ostralegus*) and shelduck (*Tadorna tadorna*) had some of the highest concentrations of ^{137}Cs of all the birds in their tissues, yet their breeding success and populations were unaffected.

Black-headed gulls, on the other hand, were found to be feeding mainly inland, and were the least contaminated with radionuclides of all the birds at Ravenglass, yet this species and its breeding success were in decline. Calculations of the total dose equivalent rate to the whole body of the most contaminated black-headed gull amounted to 9.8×10^{-4} msv h^{-1} (about 8.4×10^{-4} mGy h^{-1}, whole-body absorbed dose rate), and the background exposure dose was of the order of 8.3×10^{-4} mGy h^{-1}. As a minimum chronic dose of 1000 mGy day^{-1} has been found necessary to retard growth of nestling birds, and 9600 mGy over 20 days of incubation to cause the death of 50% of embryos in black-headed gulls' eggs in laboratory experiments, the concentrations of radionuclides in the food, body tissues and general environment were at least three orders of magnitude too low to have any effect.

The more likely cause of the desertion of the gullery was the combination of an uncontrolled fox population, the severest outbreak of myxomatosis among the rabbits since 1954, and the driest May–July period on record, all in the same year (1984).

Of growing interest are 'mesocosms' – small, carefully controlled systems which provide some simulation of the conditions in the natural environment. Experimental ponds provide an example of mesocosms. Ponds of standard size can be established, which become colonized by plants, insects and vertebrates. The effects of chemicals on the pond communities can then be tested. Such mesocosms provide a halfway house between closely controlled laboratory tests and extensive, only loosely controlled full-scale field trials. One of their advantages is that they do allow replication, which is not usually possible in full-scale field trials. (For reviews of mesocosms see Crossland (1994) and Ramade (1992).)

Further Reading

CALOW, P (ed.) (1993, 1994) *Handbook of Ecotoxicology*, Vols 1 and 2 A comprehensive text, describing all the main types of ecotoxicity testing and the scientific basis for them.

LEWIS, M. A. (1990) A review of algal toxicity testing.

TIMBRELL, J. (1995) *Introduction to Toxicology*, 2nd Edition A straightforward account of toxicity and toxicity testing.

7

Biochemical Effects of Pollutants

7.1 Introduction

When pollutants enter living organisms they cause a variety of changes (for a general account, see Guthrie and Perry (1980), Timbrell (1995) and Hodgson and Levi (1993)). These changes (bio-effects) are, broadly speaking, of two kinds: those which serve to protect the organisms against the harmful effects of the chemical, and those which do not. Some examples of both types are given in Table 7.1.

We consider, first, *protective responses*. Some protective mechanisms work by reducing the concentration of free pollutants in the cell, thereby preventing or limiting interactions with cellular components which may be harmful to the organism. Organic pollutants often cause the induction of enzymes that can metabolize them (see also Chapter 5). One of the most important of these enzyme systems is the monooxygenase system whose function is to increase

Table 7.1 Protective and non-protective responses to chemicals

Type of effect	Example	Consequences
Protective	Induction of monooxygenases	Increase in rate of metabolism of pollutant to more water-soluble metabolite and thus increase in rate of excretion (Chapter 5)
	Induction of metallothionein	Increases the rate of binding with metals to decrease bioavailability
Non-protective (may or may not lead to toxic manifestations)	Inhibition of AChE	Toxic effects seen above 50% inhibition
	Formation of DNA adducts	May cause harmful effects if leading to mutation

the rate of production of water-soluble metabolites and conjugates of low toxicity which can be rapidly excreted. In this case metabolism causes detoxification. However, in a small, yet important, number of cases, metabolism leads to the production of active metabolites (e.g. of carcinogens or organophosphorus insecticides) that can cause more damage to the cell than the original compounds. An outline of these reactions is shown in Figure 7.1.

Another mechanism whereby the bioavailability of pollutants is reduced is by binding to another molecule. This can lead to excretion or storage. The metallothioneins are examples of proteins which can bind metal ions and which become induced when there is exposure to high metal concentrations. There are also inducible proteins which can bind organic pollutants; this mechanism provides the basis of resistance to some drugs by removing them from the cell.

In addition to protective mechanisms which control the levels of free pollutants, there are other responses which are concerned with the repair of damage caused by pollutants. The release of stress proteins falls into this category. When organisms are exposed to chemical insult or heat shock, stress proteins are released which have the function of repairing cellular damage. Similarly, if pollutants cause damage to DNA, repair mechanisms come into play (Figure 7.2). There are many examples of homeostatic mechanisms like these which restore cellular systems to their normal functional state after toxic damage has been caused by chemicals.

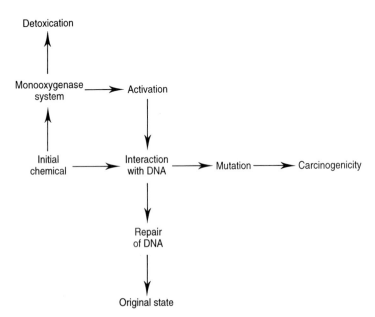

Figure 7.1 Pathways for activation and detoxication of chemicals.

system that usually protects the organism) can sometimes generate the reactive compounds that cause cellular damage. This phenomenon is considered in more detail in section 7.4.1 on genotoxic compounds.

Toxic effects are sometimes due to molecular interactions which do not lead to the formation of covalent bonds. The reversible binding of a molecule to a site on a cellular macromolecule can lead to toxicity. The target may be a receptor site for a chemical messenger (e.g. a receptor for acetylcholine, or GABA (gamma aminobutyric acid) or it may be a pore through a membrane which normally allows the passage of ions (e.g. the blockage of Na^+ channels by tetrodotoxin).

To gain a full understanding of the toxic effects of chemicals, it is necessary to link initial molecular interactions to consequent effects at higher levels of organization. The extent to which such a molecular interaction occurs is, in a general way, related to the dose of the chemical received – although the relationship is rarely a simple one. There are several reasons why many dose–response relationships are not simple. Low-level exposure, up to a particular threshold, may produce no measurable interaction at all. In some cases protective mechanisms remove the chemical before it can reach its site of action. For this reason there are often major differences between *in vitro* and *in vivo* experiments. In other cases when the dose exceeds a particular threshold level, protective mechanisms may come into play which reduce the amount of chemical reaching its active site. An example of this is the induction of metallothionein which decreases the bioavailability of toxic metals such as cadmium. Another reason for lack of effect at low levels of exposure is that many systems have reserve capacity. The enzyme carbonic anhydrase, which catalyzes the conversion of carbon dioxide to carbonic acid, has to be inhibited by more than 50% before physiological effects are seen. Similarly, the inhibition of brain acetylcholinesterase must usually exceed 50% before consequent physiological disturbances are seen (Table 7.1). In other cases, especially carcinogenicity, it has been argued that no minimum safe level exists. In theory, a single molecular interaction can initiate the whole process leading to the development of cancer.

Thus, dose–response relationships may be complex in regard to the molecular interactions which underlie toxicity. Clearly there can be further complications when relating dose to toxic manifestations at higher levels of organization (e.g. toxic effects seen at the level of cell function, or the function of the whole organism). Here, it is not just a question of the relationship between dose and the degree of molecular interaction (e.g. the percentage inhibition of an enzyme); there may also be a complex relationship between the degree of molecular interaction and consequent higher level effects – e.g. due to the intervention of protective mechanisms such as the induction of stress proteins. In studying responses to toxic molecules it is very important to construct appropriate dose–response curves over the range of exposures that are likely to be experienced. It would be unwise to assume that a dose–response curve is a simple straight-line relationship.

7.4 Examples of Molecular Mechanisms of Toxicity

Molecular interactions between xenobiotics and sites of action, which lead to toxic manifestations, may be highly specific for certain types of organism or very non-specific. In the simplest case, molecules are highly selective between two species because one species has a certain type of site of action which does not occur in the other. For example, the pesticide dimilin acts on the site of formation of chitin and thus affects only those arthropods that utilize this material to form their exoskeleton. The organophosphorus insecticides which act on the nervous system are toxic to all animals but have little or no toxicity towards plants. Animals have a site of action in the nervous system (in the example given, a form of acetylcholinesterase, discussed in more detail in the section on neurotoxic compounds, 7.4.2.) whereas plants have no nervous systems and no comparable site of action. Some compounds show little selective toxicity and may be regarded as general biocides. An example is dinitro-orthocresol (DNOC) and related compounds which act upon mitochondrial membranes, and cause the uncoupling of oxidative phosphorylation. This system is found in all eukaryotes, and 'uncouplers' of this type can run down the gradient of protons across mitochondrial membranes in general, thereby inhibiting or preventing the synthesis of ATP. (Further discussion of these compounds is given in section 7.3.1.3.)

Some of most subtle selectivity occurs in resistant strains, a subject dealt with in more detail in Chapter 12. Here two strains of the same species can have different forms of the same site of action – one of them susceptible to a toxic molecule, the other non-susceptible, e.g. some strains of insects resistant to organophosphorus insecticide have forms of acetylcholinesterase which differ from those of susceptible strains of the same species. The difference between the 'susceptible' and 'resistant' forms of acetylcholinesterase may be just a single amino acid in the entire molecule – and thus a very small difference in the structure of a site of action can cause a large difference in toxicity. An organophosphorus insecticide may show little tendency to interact with the 'resistant' enzyme.

In the following account, animals and plants will be considered separately.

7.4.1 Genotoxic Compounds

Many compounds which act as carcinogens are known to cause damage to DNA (i.e. they are genotoxic), and it is strongly suspected that this is a causal relationship. When cells with damaged DNA divide, mutant cells can be produced. Some mutant cells are tumour cells which will follow uncoordinated growth patterns, and may migrate within the organisms to produce secondary growths (metastasis) in other locations.

The relationship between DNA changes and harm to the organism is complex. While adduct formation (covalent binding of the pollutant to DNA) is a good index of exposure, the relationship of adduct formation to harm to the organism is less well defined. For the most part, these DNA adducts are short lived. DNA-repair mechanisms quickly excise the adducted structures and replace them with the original moiety (Figure 7.2). Sometimes, however, they are relatively stable, and when the cell divides, the adducted element is mistranslated, and a mutant cell is produced. Thus, while there are good data to relate the number of DNA–BaP adducts and extent of cigarette smoking, the relationship between DNA–BaP adducts and lung cancer is less well established.

Certain polycyclic aromatic hydrocarbons or 'PAHs' (e.g. BaP, dibenz(ah)anthracene), acetylaminofluorene, aflatoxin, and vinyl chloride are all examples of genotoxic pollutants. In all of them, it is not the original compound that interacts with DNA. Indeed the original compounds are relatively stable and unreactive. Enzymatic metabolism (mainly oxidative metabolism by one or more forms of monooxygenase) produces highly reactive and short-lived metabolites which can bind to DNA (Figure 7.1).

This type of molecular interaction would appear to be common and widespread. A significant number of pollutants are known to have mutagenic properties – certain PAHs being a case in point. PAHs are present in smoke and soot, and in crude oil. They are, therefore, present in urban areas, wherever there is traffic, smoke and soot, and in marine locations where there have been oil spillages.

7.4.2 *Neurotoxic Compounds*

The nervous system of both vertebrates and invertebrates is very sensitive to the toxic effects of chemicals and there are many examples of neurotoxins – both naturally-occurring and man-made. Among the 'natural' neurotoxins should be included tetrodotoxin (from the puffer fish), botulinum toxin (from the anaerobic bacterium *Clostridium botulinum*), atropine (from the deadly nightshade, *Atropa belladonna*), the natural insecticides nicotine (from the wild tobacco *Nicotiana tabacum*), pyrethrin (from the flowering heads of *Chrysanthemum* sp) and many more. Among man-made compounds it is interesting to note that the four major groups of insecticide – organochlorine, organophosphorus, carbamates and pyrethroid insecticides – all act as nerve poisons.

All of the examples given disturb, in some way, the normal transmission of impulses along nerves and/or across synapses (i.e. junctions between nerves or between nerve endings and muscle and gland cells). However, a distinction can be made between compounds which act directly upon receptors or pores situated in the nerve membrane, and those which inhibit the acetylcholinesterase (AChE) of synapses. These two groups will now be considered separately.

The passage of an action potential along a nerve is dependent upon the flow of Na^+ and K^+ across the nerve membrane. During the normal passage

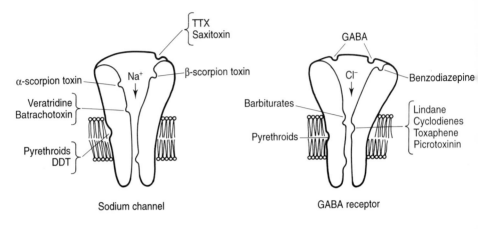

Figure 7.4 Sodium channels and GABA receptors.

of an action potential, Na$^+$ channels (Figure 7.4) are open for a brief instant, allowing the inward flow of Na$^+$ ions. They are then closed to terminate the Na$^+$ flow. Pyrethroid insecticides, natural pyrethrins and DDT all interact with Na$^+$ channels to disturb this function. The usual consequence of their interaction is retarded closure of the channels. This can cause a prolongation of the flow of Na$^+$ ions across the membrane, which leads to disturbance of the normal passage of the action potential. This type of poisoning causes uncontrolled repetitive spontaneous discharges along the nerve. Several action potentials are generated in response to a single stimulus instead of just one. Uncoordinated muscular tremors and twitches are characteristic symptoms of this type of poisoning. The interaction of these hydrophobic 'water-insoluble' compounds with Na$^+$ channels is reversible and does not appear to involve the formation of covalent bonds. It is likely that these compounds first dissolve in the lipids of the nerve membrane before interacting with some site on the Na$^+$ channel, which spans the membrane.

Another site of action for insecticides and other neurotoxins is the GABA receptor, which functions as a Cl$^-$ channel through the nerve membrane (Figure 7.4). Chlorinated cyclodiene insecticides, or their active metabolites (e.g. dieldrin, endrin, heptachlor epoxide), act as GABA antagonists. By binding to the receptors they reduce the flow of Cl$^-$ ions. So, too, does the insecticide γHCH (lindane) and naturally occurring picrotoxinin. In vertebrates these compounds act upon GABA receptors of the brain. Convulsions are typical symptoms of this type of poisoning.

Receptors for acetylcholine, which are located on post-synaptic membranes (i.e. on the other side of the synapse from nerve endings), represent the site of action for a number of chemicals. Thus, nicotine can act as a partial mimic (agonist) of acetylcholine at what are termed 'nicotinic receptors' for acetylcholine. Similarly, atropine can act upon 'muscarinic' receptors for acetyl-choline. Nicotinic receptors and muscarinic receptors differ from one another in their structure, their response to toxic chemicals and their location within the nervous system. In vertebrates nicotinic receptors are found especially at the neuromuscular junction, and on many synapses of autonomic ganglia. By contrast, muscarinic receptors are especially found on synapses at the endings of parasympathetic nerves – typically on the membranes of smooth muscle or gland.

Compounds that inhibit AChE represent one of the most toxic groups of compounds to vertebrates and invertebrates. As noted earlier (Chapter 1), OPs are particularly important here. Some were developed for chemical warfare, many others have been developed as insecticides. Another group of anticholin-esterases are the carbamates which have insecticidal action. (These should not be confused with other carbamates used as herbicides or fungicides, which do not act as anticholinesterases.)

The toxic action of these compounds is indirect, in that their primary effect is to inhibit the action of AChEs which have the function of breaking down acetylcholine released into the synapse. Acetylcholine release occurs from the endings of cholinergic nerves and it functions as a chemical messenger. When an impulse reaches a nerve ending, acetylcholine is released and carries the signal across the synaptic cleft to a receptor on the post-synaptic membrane (which may be of a nerve cell, a muscle cell or a gland cell). When acetyl-choline interacts with its receptor, a signal is generated on the post-synaptic membrane, so that the impulse (message) is carried on. For effective neuronal control, it is essential that this signal is rapidly terminated. To achieve this, acetylcholine must be rapidly broken down by AChE in the vicinity of the receptor (Figure 7.3). Anticholinesterases have the effect of reducing or pre-venting altogether the breakdown of acetylcholine. As a consequence, acetyl-choline builds up in the synapse, leading to 'overstimulation' of the receptor and the continued production of a signal after this should have stopped. If this situation continues, the signalling system will eventually run down, resulting in synaptic block. At this point, it will no longer be possible for acetylcholine to relay signals across the synapse. In the case of neuro-muscle junctions so affected, tetanus will result, with the muscle in a fixed state, unable to contract or relax in response to nerve stimulation.

7.4.3 *Mitochondrial Poisons*

Mitochondria have a vital role in energy transformation and are found in all eukaryotes. It is not, therefore, surprising, that some of the most dangerous non-selective biocides act upon mitochondrial systems.

Uncouplers of oxidative phosphorylation such as 2,4-dinitrophenol, fall into this category. When a mitochondrion is functioning normally, it has a gradient (electrochemical gradient) of proteins across its inner membrane

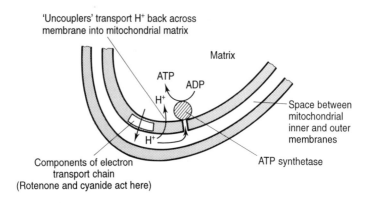

'Uncouplers' transport H⁺ back across membrane into mitochondrial matrix

Figure 7.5 Mitochondrial poisons. The figure shows a diagrammatic cross-section of a mitochondrion. Protons (H^+) are actively transported from the matrix, across the inner membrane, as a result of electron flow along the electron transport chain. These protons can then flow back to the matrix via the enzyme ATP synthetase, where energy associated with them is used to synthesize ATP. 'Uncouplers' such as 2,4-dinitrophenol, can carry these protons back to the matrix before ATP synthesis occurs. Other poisons, e.g. rotenone, CN^-, can inhibit the flow of electrons along the electron transport chain.

(Figure 7.5). In fact there is an excess of protons on the outside of the inner membrane and a deficiency on the inside. The maintenance of this gradient depends upon the mitochondrial membrane being essentially impermeable to protons. The production of ATP by mitochondria is driven by energy stored in this proton gradient. If the proton gradient is lost, ATP production will cease. The compounds in question can eliminate this gradient, dissipating in the form of heat the energy associated with it and thus preventing the energy stored in a proton gradient from being used to biosynthesize ATP. Other mitochondrial poisons such as the naturally occurring insecticide rotenone, and cyanide ions, can inhibit the operation of the electron transport chain of the inner membrane of the mitochondrion, thus preventing the production of the proton gradient referred to above. They do so by interacting with components of the electron-carrier system.

7.4.4 *Vitamin K Antagonists*

Warfarin and certain related rodenticides are toxic to vertebrates because they act as antagonists of vitamin K. Vitamin K plays an essential role in the synthesis of clotting proteins in the liver. It undergoes a cyclical series of changes (vitamin K cycle) during the course of which the clotting proteins become carboxylated. After carboxylation has occurred, the clotting proteins are released into the blood, where they play a vital role in the process of clotting which occurs when there is damage to blood vessels. Warfarin and related compounds have a structural resemblance to vitamin K, and strongly compete with it for binding sites, even at low concentrations. This leads to inhibition of the vitamin K cycle, and consequently to the incomplete synthesis of clotting proteins. Under these conditions the levels of clotting proteins fall in blood, and the blood loses its ability to clot. Failure of clotting results in haemorrhage.

Flocoumafen and related compounds, referred to as second-generation anticoagulant rodenticides, were developed to get round the problem of resistance to warfarin. These rodenticides become lethal when the receptor sites in the liver are saturated. There is a dramatic difference in the toxicity of flocoumafen to rats and to quail (LD_{50} 0.25 mg kg^{-1} and 300 mg kg^{-1} respectively). An important factor in this species difference is that the rodenticide is metabolized more rapidly by quail than by rat (Huckle *et al.*, 1989).

7.4.5 Thyroxine Antagonists

The thyroid gland produces two hormones which have marked effects upon metabolic processes in many tissues. One of these, thyroxine (1-3,3,5,5-tetra-iodo thyronine or T4), binds to the protein transthyretin (TTR) which is part of a transport–protein complex found in blood. The other part of the complex consists of a second protein to which is bound retinol (vitamin A) (Figure 7.6). Thus, transthyretin : thyroxine is attached to the retinol-binding protein (RBP). Certain hydroxy metabolites of the PCB congener 3,3'-4,4'-tetra-chlorobiphenyl (3,3,'-4,4'-TCB) compete with thyroxine for its binding site on TTR. In particular, one metabolite 4'-hydroxy-3,3',4-5'tetrachlorobiphenyl binds very strongly and effectively competes with thyroxine (Brouwer *et al.*, 1990). These metabolites are formed due to metabolism of the original PCBs congener by a particular cytochrome P_{450} form of the monooxygenase system (P_{450} IAI).

When binding occurs there are two consequences. First, thyroxine is displaced, and lost from the blood. Second, the retinol complex breaks away, and retinol is lost from the blood. Thus, the levels of thyroxine and retinol will fall as a consequence of the production of certain PCB metabolites.

7.4.6 Inhibition of ATPases

The ATPases (adenosine triphosphatases) are a family of enzymes ($Na^+K^+ATPase$, Ca^{2+}-ATPase etc.) involved in the transport of ions. These enzymes are involved in the osmoregulation of a variety of organisms and the effects of a number of organochlorines of this process have been investigated. The avian salt gland which enables pelagic seabirds to maintain their salt balance in a marine environment is also dependent on ATPases. Another ATPase-dependent organ is the avian oviduct. The inhibition of Ca^{2+}-

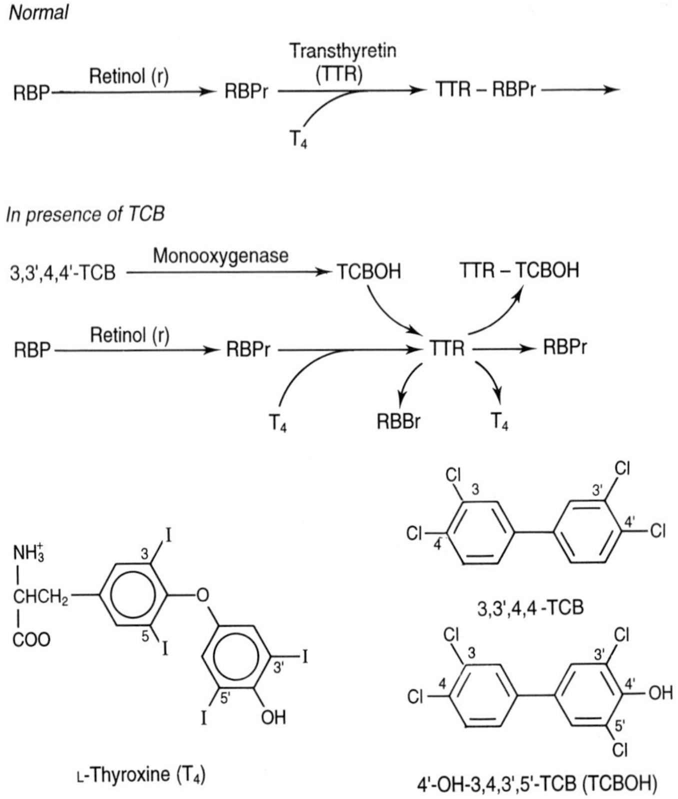

Normal

In presence of TCB

L-Thyroxine (T₄)

3,3',4,4-TCB

4'-OH-3,4,3',5'-TCB (TCBOH)

Figure 7.6 Mechanism of toxicity of a polychlorinated biphenyl. Retinol (r) binds to retinol-binding protein (RBP) which is then attached to transthyretin (TTR). Thyroxine (T4) binds to TTR, and is transported via the blood in this form. The coplanar PCB, 3,3',4,4'-tetrachlorobiphenyl (3,3',4,4'-TCB) is converted into hydroxymetabolites by the inducible cytochrome P_{450} called P_{450} 1A1. The metabolite 4'-OH-3,3',4,5'-tetrachlorobiphenyl (TCBOH) is structurally similar to thyroxine, and strongly competes for thyroxine binding sites. The consequences are loss of thyroxine from TTR, the fragmentation of the TTR–RBPr complex, and loss of both thyroxine and retinol from blood. (After Brouner et. al. (1990)).

ATPase, which is involved in the transport of calcium, by DDE (the persistent metabolite of DDT), is considered to be the basis of DDE-induced eggshell thinning. This phenomenon is considered in some detail in Chapter 15.

7.4.7 *Environmental Oestrogens and Androgens*

There has been considerable concern recently about the present of man-made oestrogenic and androgenic chemicals in the environment. The best documented environmental effect is the induction of vitellogenin in male fish.

Vitellogenin is a protein synthesized in the liver of female fish and transported to oocytes to form the yolk of the eggs. It is well established that the induction of vitellogenin in the female fish is under oestrogenic control. A survey of fish exposed to sewage treatment effluent in the UK was carried out in 1988 as a consequence of the observation by anglers of hermaphrodite fish in sewage treatment water lagoons (Purdom *et al.*, 1994). High levels of vitellogenin were found in male fish in most cases although the phenomenon was not observed in rivers or reservoirs. The nature of the chemical (or chemicals) causing this effect has not yet been firmly established but the nonylphenols, widely used in detergents, are considered the most likely cause.

Pollutants with androgenic activity can cause masculinization of female gastropods. Imposex (females developing male characteristics such as penis and vas deferens) has been widely reported in marine gastropods associated with marinas. Detailed studies have been carried out at the Plymouth Marine Laboratory since the first finding of imposex in the dog whelk (*Nucella lapillus*) in Plymouth Sound in 1969. Laboratory experiments and *in situ* transfer experiments have shown that imposex is initiated in dog whelks by tributyl tin at very low concentrations. This phenomenon is considered in more detail in Chapter 15.

In both this and the preceding example the underlying cause of change is considered to be that chemicals mimic the action of true hormones at the receptor site and thus trigger the reactions that would normally be caused by the natural hormone. The problem is caused by the fact that these unnatural 'hormones' trigger reactions in the wrong sex.

Another series of problems can be caused when chemicals block the hormone receptor site. In this case the normal action of the hormone is inhibited since it cannot react with the receptor. (An example of this mechanism is the action of tetrachlorodibenzodioxin (TCDD).) The means of assessing this compound and other dioxins and related chemicals is considered in Chapter 12.

7.4.8 Reactions with Protein Sulphydryl (SH) Groups

The ions Hg^{2+} and Cd^{2+} are toxic to many animals. The main reason for this appears to be their ability to combine with sulphydryl (thiol) groups, thereby preventing normal function. Sulphydryl groups on enzymes and other proteins have important functional roles, e.g. the formation of disulphide bridges and consequent conformational changes in the proteins. With this kind of toxic interaction it may be difficult to establish which sulphydryl groups on which proteins represent the sites of toxic action.

Organomercury is more lipophilic than inorganic mercury, and has a different distribution in the body. It tends to move into fatty tissues and cross membranes, including those of the blood–brain barrier. As a consequence, the toxicity of organomercurial compounds tends to be expressed in the brain, while that of inorganic mercury is expressed in peripheral tissues. Other organometallic compounds, e.g. lead tetraalkyl, also tend to have their toxic action in the brain.

7.4.9 *Photosystems of Plants*

A number of herbicides which show little toxicity to vertebrates or insects, act as poisons of the photosynthetic systems of plants. Substituted ureas and triazines are examples. By mechanisms that are not yet clear, they interrupt the flow of electrons through the photosystems that are responsible for the light-dependent reaction of photosynthesis – i.e. the splitting of water to release molecular oxygen.

7.4.10 *Plant Growth Regulator Herbicides*

A group of herbicides, sometimes called the phenoxyalkanoic herbicides, have growth-regulating properties. MCPA, 2,4D, CMPP, and 2,4DB are well known examples. The molecular mechanism by which they affect growth has never been clearly established, and the site at which they act has not been defined. However, it is interesting that exposure of plants to them causes the production of ethylene, itself a growth-regulating compound.

These herbicides have very low toxicity to vertebrates and insects which, it may be presumed, do not have receptors for growth regulation comparable to those of plants.

Further Reading

TIMBRELL, J. A. (1992) *Principles of Biochemical Toxicology*, 2nd Edition A long chapter is devoted to biochemical mechanisms of toxicity of importance in medical toxicology.

HODGSON, E. and LEVI, P. (1993) *Introduction to Biochemical Toxicology*, 2nd Edition Many examples given of mechanism of action of toxicants, including most of those mentioned in this chapter.

HASSALL, K. A. (1990) *The Biochemistry and Uses of Pesticides*, 2nd Edition Describes the mode of action of many pesticides.

HODGSON, E. and KUHR, R. J. (1990) *Safer Insecticides: Development and Use* Detailed account of the mode of action of certain insecticides.

8

Physiological Effects of Pollutants

8.1 Introduction

Pollutants may damage organisms with lethal consequences (as described in Chapters 6 and 7). The effect on the population is then an increase in the mortality rate of at least one age class. Alternatively there may be damage to, or effects on, the machinery of reproduction, resource acquisition and uptake. These effects are described in detail in this chapter. Where resource uptake is reduced, there are consequent reductions of birth rate and/or somatic growth rate (here referred to jointly as 'production rate'), and these depress population growth rate.

Detoxication mechanisms generally use resources, including energy, and these resources are consequently not available for production. Thus, production is likely to be reduced when detoxication occurs. The overall effects of detoxication mechanisms on production and mortality are considered at the end of this chapter. Note at this stage, however, that either damage or detoxication may result in reduced production in a polluted environment.

With these points in mind, the effects of pollutants on organisms will be considered at several different levels of organisation (Figure 8.1). In this chapter, we are concerned mainly with physiological effects above the biochemical level (covered in Chapter 7). Physiology is defined as the branch of science concerned with the functioning of organisms, and the processes and functions of all or part of an organism. Thus, in ecotoxicology, we are concerned with describing the disruptive effects of pollutants on 'normal' physiology, normal being the stresses and conditions to which an organism would be exposed in the absence of pollution. These stresses may be completely novel and occur in response to man-made chemicals that have appeared in the environment recently (on the geological time-scale), or they may simply be an increase in a response to a substance to which the organism has evolved natural protection mechanisms (e.g. metals).

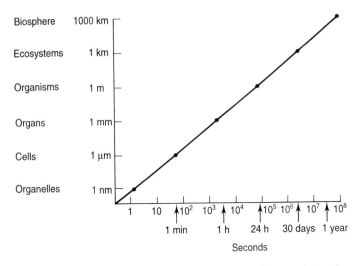

Figure 8.1 Schematic graph of the relationship between complexity and size of natural systems and 'compartments' (= 'black boxes') and typical response times to pollutant 'insults'. After various authors.

8.2 Effects of Pollutants at the Cellular Level

When a pollutant enters a cell it may trigger certain biochemical responses which have evolved either to break the chemical down (Chapter 7), or store it in such a way that it is 'hidden' within a compartment preventing interference with essential biochemical reactions within the cell. For example, in invertebrates there are clearly defined pathways for metal detoxication (Dallinger, 1993; Hopkin, 1989, 1990). Perhaps the simplest case is the epithelium of the digestive system of terrestrial invertebrates which is usually only one cell in thickness and acts as a barrier between the internal environment of the animal (i.e. the blood bathing the organs) and the food in the lumen. Therefore, storage mechanisms and/or methods of exclusion have to be extremely efficient because terrestrial invertebrates (unlike aquatic organisms) are not able to excrete xenobiotics from the blood into the external medium across the respiratory surfaces if they are taken up to excess. In land mammals, lipophilic organics are converted to water-soluble metabolites and conjugates which are then excreted in the bile and urine (see Chapter 5).

Three main detoxication pathways have evolved for the binding of metals which enter the epithelial cells (Figures 8.2, 8.3). Although the chemistry of binding appears to be similar in all terrestrial invertebrates so far examined, the subsequent fate of the waste material is controlled by the digestive processes of the animals in question. An example of these differences is provided by the contrasting ways in which metals are bound in the hepatopancreas of

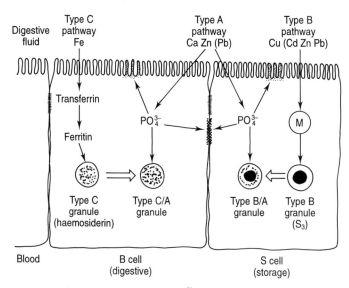

Figure 8.2 Schematic diagram showing the three pathways of detoxication of metals from the digestive fluids by the B and S cells of the hepatopancreas of the woodlouse, *Porcellio scaber*. Reproduced from Hopkin (1990) with permission of the British Ecological Society.

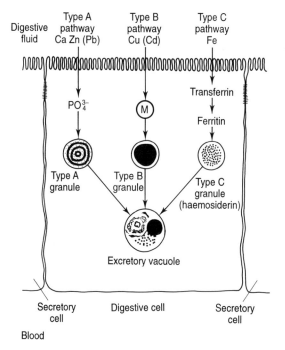

Figure 8.3 Schematic diagram showing the three pathways of detoxication of metals from the digestive fluids by the digestive cells in the hepatopancreas of the woodlouse-eating spider, *Dysdera crocata*. Reproduced from Hopkin (1990) with permission of the British Ecological Society.

both the terrestrial isopod *Porcellio scaber* and a major predator of isopods, the woodlouse-eating spider *Dysdera crocata* (Figure 8.4, see Box 8.1).

8.3 Effects at the organ level

When pollutants are ingested by organisms, or pass into the blood across respiratory epithelia or the external surface of the body, they may be 'compartmentalized' in particular organs within the body. This may be accidental or deliberate. For example, radioactive isotopes of essential elements will travel to the same locations in tissues as their non-radioactive counterparts. Radioactive iodine accumulates in the thyroid gland of mammals and may cause thyroid cancer. Cadmium is accumulated in the liver and kidneys of mammals. The symptoms of cadmium poisoning are proteinuria (excretion of proteins in the urine) due to the breakdown of the kidney cells when cadmium levels exceed a 'critical concentration'. This concept of the critical target organ concentration is very important in ecotoxicology. With small organisms, analysis of concentrations of pollutants in whole individuals may be misleading if the contaminant is localized strongly within a specific part of the body. For example, over 90% of the cadmium, copper, lead and zinc in woodlice from metal-contaminated sites is contained within the hepatopancreas, an organ which comprises less than 10% of the weight of the whole animal (Hopkin, 1989). In regions contaminated heavily with metals, concentrations of zinc can exceed 2% of the dry weight of the hepatopancreas of woodlice. In these circumstances, the detoxification capacity of the organ is eventually exceeded, the cells begin to break down and the woodlice becomes moribund due to zinc poisoning (Figure 8.6).

8.4 Effects at the Whole-Organism Level

8.4.1 *Effects on Behaviour of Aquatic Animals*

The effects of pollution on behaviour have been reviewed with primary reference to aquatic animals by Atchison, Sandheinrich and Bryan (1996) which should be consulted for further information on the material in this section. While all behaviours are potentially vulnerable to alteration by pollutants, most work has been done on foraging, with some attention to vigilance. Impaired foraging results in reduced resource acquisition and so in reduced production. Impaired vigilance results in increased vulnerability to predators and so to increased mortality rate. In these ways the effects of pollution on behaviour may result in lowered production and increased mortality rate.

The components of foraging behaviour are illustrated diagrammatically in Figure 8.7. All these components may be adversely affected by pollution. Thus, although little work has been done on the effects on appetite, a common con-

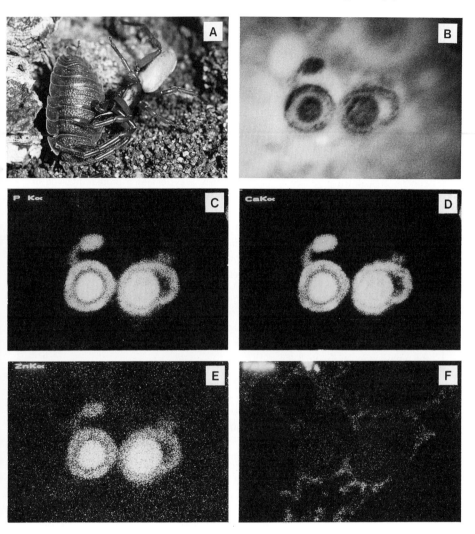

Figure 8.4 (A) *Dysdera crocata* attacking a specimen of *Porcellio scaber* of 1 cm in length. (B) Scanning transmission electron micrograph (bright-field image) of an unstained resin-embedded section (0.5 μm in thickness) through two type A calcium phosphate granules in a digestive cell of the hepatopancreas of *D. crocata*. Diameter of each granule section = 1 μm. (C–F) Electron-generated X-ray maps for phosphorus (C), calcium (D), zinc (E) and iron (F) of the type A granules shown in (B). Philips CM12 STEM, EDAX 9900 X-ray analyzer, screen resolution 256–200 pixels, dwell time on each pixel 50 ms, spot diameter 20 nm. Reproduced from Hopkin (1990) with permission of the British Ecological Society.

sequence of chemical stressors is the cessation of feeding. Prey encounter rates depend on many factors including search strategy, learning and sensory systems. All can be affected by chemical stressors, reducing the efficiency of searching for prey. Little work has been done on prey choice, but a few studies

Box 8.1

The **type A pathway** is involved in the intracellular precipitation of calcium and magnesium as phosphates. In the hepatopancreas of *Porcellio scaber* (Figure 8.2), zinc and lead may be present in this type A phosphate-rich material which is deposited on the cytoplasmic side of the cell membranes, and around existing metal-containing granules. However, in *Dysdera crocata* (Figure 8.3) the type A material forms discrete granules with a characteristic concentric arrangement of layers in thin section (Figures 8.4B, 8.5). Zinc (Figure 8.4E) and lead are also found associated with these granules.

The **type B pathway** is followed by metals such as copper and cadmium which have an affinity for sulphur-bearing ligands. The sulphur-rich type B granules probably contain breakdown products of metallothionein, a cysteine-rich protein involved in the intracellular binding of zinc, copper, cadmium and mercury, and have their origin in the lysosomal system (Dallinger, 1993). Lead may also occur in type B granules. In isopods (but not in spiders), some type B granules may be surrounded by a layer of type A material (type B/A granules following the convention that the type of material first accumulated in the granule is given precedence) but granules with a type A core surrounded by type B material have not yet been discovered.

The **type C pathway** is exclusively for the accumulation of waste iron in isopods and spiders. Type C granules are probably composed of haemosiderin, a breakdown product of ferritin. In isopods, type A material may be mixed with the iron-rich type C material to form type C/A granules (Figure 8.2). In *Dysdera crocata*, iron is not found in the type A granules (Figure 8.4F).

Once the type A, B and C material has been precipitated, there is no evidence that it is remobilized. Indeed, the only route by which the granules can be excreted is by voiding of the contents of the cell into the lumen of the digestive system for subsequent excretion in the faeces. The granules therefore represent a storage detoxication system.

In isopods, type A material occurs in both cell types of the hepatopancreas but type B and C material is restricted to the S and B cells respectively. Large numbers of B cells break down during each 24-h digestive cycle but S cells are permanent and never void their contents into the lumen of the hepatopancreas. Thus, material deposited in the S cells remains there until the isopod dies. In contrast, the spider stores all three types of material in a single cell type, the digestive cell (Figure 8.3). Large numbers of these cells break down at the end of each digestive cycle and void their contents. This waste contains large numbers of type A, B and C granules which are excreted subsequently in the faeces (Hopkin, 1989).

Thus, terrestrial isopods accumulate metals in the S cells throughout their lives so that in contaminated sites, concentrations in the hepatopancreas reach very high levels. In contrast, the spider regulates the concentrations in the hepatopancreas by excreting metals assimilated in the digestive cells. Consequently, concentrations of zinc, cadmium, lead and copper do not deviate from normal over the long term in the hepatopancreas of *Dysdera crocata*, even when the spiders are fed on heavily contaminated woodlice for an extended period.

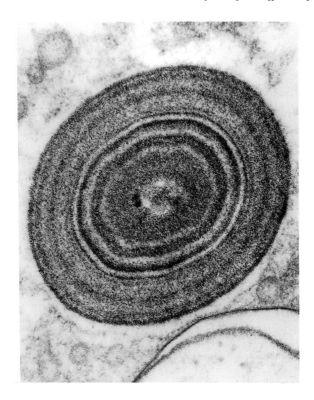

Figure 8.5 Concentrically-structured type A granule from the hepatopancreas of the woodlouse-eating spider *Dysdera crocata* (diameter 2 μm). Reproduced from Hopkin (1989) with permission of Elsevier Applied Science.

show that it, too, can be affected. Most importantly, capture rates are known to be affected by toxicants in some species, at least when larger or more evasive prey are being hunted. The time from capture to ingestion ('handling time') has often been shown to be increased by toxicants, for example by copper in bluegill feeding on *Daphnia*, by zinc and lead in zebrafish (*Brachydanio rerio*), and by alkyl benzene sulphonate detergent in flagfish (*Jordanella floridae*) (Atchison *et al.*, 1996). However, these increases in handling times seem to be due to repetitive rejection and recapture, perhaps caused by blockage of the gustatory senses by the contaminant. These predators use vision to identify, pursue, capture and start processing their prey. Rejection may result from lack of gustatory confirmation that the captured item is edible.

Foraging is not the only goal, however, since as mentioned above an animal must also avoid itself being eaten. 'Few failures in life are as unforgiving as the failure to avoid predation' (Lima and Dill, 1990). Most studies of vulnerability to predation have been simple experiments in which prey were

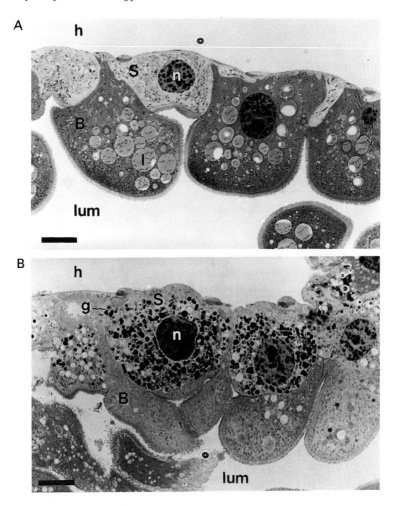

Figure 8.6 Light-micrographs of B cells (B) and S cells (S) within the hepatopancreas of two specimens of the woodlouse, *Oniscus asellus*, 6 weeks after release from the same brood pouch of a female from an uncontaminated woodland. The S cells of the juvenile reared on leaf litter contaminated with metal pollutants from a woodland near to a smelting works (B) contain far more metal-containing granules (g) than the S cells of the juvenile reared on uncontaminated litter (A). h = haemocoel; lip = lipid; lum = lumen of hepatopancreas tubule. Scale bars = 20 μm. Reproduced from Hopkin and Martin (1984) with permission of the Zoological Society of London.

first exposed to a pollutant, then to a predator. The predators were generally not exposed to the pollutant. Such studies have shown for example that ionizing radiation and mercury both increased the risk to mosquitofish (*Gambusia affinis*) of predation by largemouth bass (*Mircopterus salmoides*) (Atchison *et al.*, 1996). In other species predation risk has been shown to increase with

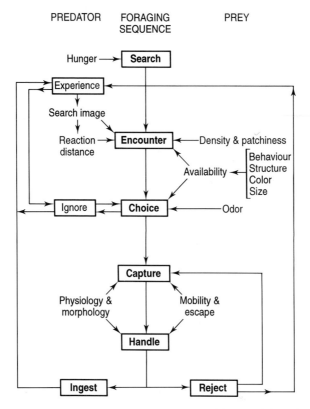

Figure 8.7 Schematic representation of the components of foraging behaviour. From Atchison *et al.* (in press).

exposure of prey to thermal stress, insecticides, pentachlorophenol, fluorene, and cadmium. Beitinger (1990) found that prey vulnerability to predation was increased in 23 of 29 experiments that he reviewed.

8.4.2 *Effects of Organophosphates on the Behaviour, Mortality and Reproductive Success of Mammals and Birds*

Organophosphorus compounds used as insecticides sometimes cause unintentional poisoning of mammals and birds. Acetylcholinesterase (AChE) activity is depressed, and this prevents normal functioning of the central and peripheral nervous systems. Behavioural responses have been documented in mammals and birds when AChE activity is depressed to less than 50% of normal levels (Grue *et al.*, 1983, 1991). It is not easy to document such effects except in blood, however, because AChE activity cannot be measured without

155

sacrificing the animal. Thus, for each animal, acetylcholinesterase activity is known at the time of death, and this is analyzed together with a behavioural profile recorded shortly before death.

Detailed quantitative analyses at low levels of toxication have only recently been attempted. Hart (1993) investigated the relationship between behaviour and AChE activity in the starling (*Sturnus vulgaris*) and found that while most behavioural effects occurred when brain levels fall below 50%, some subtle effects on behaviour, such as effects on posture (time spent standing on one leg while resting), were found when activity was in the range 50–100% of normal. Effects on posture, such as the time spent standing on one leg while resting, were found at relatively high levels of AChE activity (88% of normal), and may reflect impaired balance or coordination.

Another example of a quantitative study is shown in Figure 8.8. House sparrows (*Passer domesticus*) were dosed with an organophosphate, chlorfenrinphos, and their feeding behaviour was then recorded during four successive 1.5-h periods. The percentage seeds dropped is plotted against dose (assayed at the end of the day) in Figure 8.8. Birds whose brain AChE activity was most reduced, initially dropped some 30% more seeds than on control days (Figure

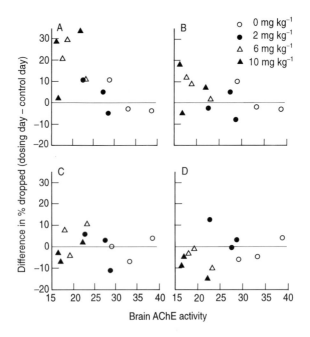

Figure 8.8 Differences in percentage seeds dropped between dosing and control days plotted against brain acetylcholinesterase activity in house sparrows. A–D show results in successive 1.5-h time bins after dosing. Doses are given in the key. From Fryday *et al.* (1994).

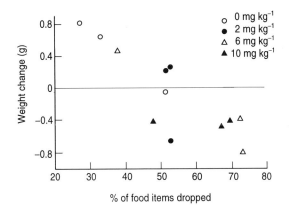

Figure 8.9 Body weight loss in relation to percentage food items dropped for the birds of Figure 8.8. From Fryday *et al.* (1994).

8.8A). However, the effect had worn off 3 h after dosing (Figure 8.8C, D). Figure 8.9 shows that the birds that dropped most seeds lost weight.

Extrapolating from the laboratory to what happens in the wild is notoriously difficult (Hart, 1993). Animals may be able to detect and avoid toxicants in their diet. If they do ingest them, they may be able to compensate for any disabilities by retreating to safe places. Alternatively, disabilities may make them more vulnerable to predation, or less able to maintain their territories or to care for offspring. In general, some field studies have succeeded but others have failed to find such effects on behaviour when brain acetylcholinesterase activity was depressed by 40–60% (Hart, 1993). This could be consistent with the existence of a general threshold for behavioural effects at about that level.

In a good example of a field study, Busby *et al.* (1990) examined the effects of the organophosphorus insecticide fenitrothion on white-throated sparrows (*Zonotrichia albicollis*) in spruce-fir forest in New Brunswick in Canada. The forest has been sprayed aerially each year since 1952 to control spruce budworm (*Choristoneura fumiferana*). Fenitrothion is the preferred insecticide. In the study of Busby *et al.* (1990), fenitrothion was applied aerially twice, once at 420 g ha^{-1} and once at 210 g ha^{-1}, 8 days later. An earlier study had shown that spraying at these levels would have resulted in brain acetycholinesterase inhibition of 42 and 30% respectively. A team of experienced field workers ringed and were able to individually identify 13 breeding pairs in the sprayed area, and 7 in a nearby control area. Each pair was followed until it abandoned its territory or its offspring fledged. The results showed that the adult population of white-throated sparrows in the sprayed area was reduced by one-third, primarily as a consequence of mortality and territory abandonment after the first spray. Other behavioural responses of breeding birds included inability to defend their territories, disruption of normal incubation patterns, and clutch desertion. Pairs which did manage to hatch at least one chick only produced one-third the normal number of fledglings. Overall the reproductive success of the pairs in the sprayed area was only one-quarter that of the pairs in the control area.

8.4.3 *Effects on Production are Measured by Scope for Growth (SFG)*

In this chapter we have seen cases in which the machinery of resource acquisition or uptake is damaged. In other cases damage is avoided, but the organism may still be affected because detoxication generally consumes energy and other resources which are therefore denied to production. Thus, either damage or detoxication is likely to lead to a loss of production.

The effect of pollution on production is usually measured by its effect on 'scope for growth' (SFG), defined as the difference between energy intake and total metabolic losses (Warren and Davis, 1967; Widdows and Donkin, 1992; Figure 8.10). An example showing the effects of tributyltin (TBT) concentration on SFG in the mussel, *Mytilus edulis*, is given in Figure 8.11. Note that above a threshold of 2 μg g^{-1}, SFG declines as TBT concentration increases, indicating a loss of production. In the field this decline could translate into a lower abundance of animals (see Chapter 12). There is here no effect of TBT on SFG at low levels of TBT. However, in the case of essential nutrients (e.g. some metals) there is a decrease in SFG at very low levels of the nutrient. SFG has been particularly useful in assessing the effects of pollution on aquatic animals (e.g. *Gammarus*, Maltby *et al.* (1990a); fish, Crossland (1988)). Field and microcosm studies have confirmed that the long-term consequences to growth and survival of individuals can be predicted from measured effects on energy balance observed at the individual level (Widdows and Donkin, 1991).

Widdows and Donkin (1991) described how reductions in SFG in *Mytilus edulis* in contaminated sites can be apportioned between specific pollutants. In a study in Bermuda, Widdows *et al.* (1990) showed that the overall reduction

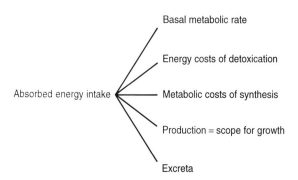

Figure 8.10 Energy/nutrient allocation diagram illustrating the definition of 'scope for growth'.

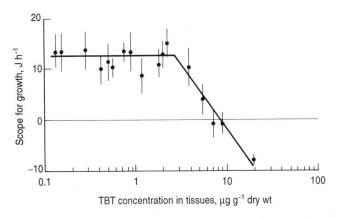

Figure 8.11 Effect of TBT (tributylin) on scope for growth (SFG) in the mussel, *Mytilus edulis*. From Widdows and Donkin (1992).

in SFG of *Mytilus edulis* could be proportional such that, at the most contaminated sites, tributyltin accounted for 21% and hydrocarbons for 74% of the observed effects.

A decline in SFG in the freshwater amphipod *Gammarus pulex* is due to increased stress; a decline in feeding rates leads to decreased offspring weight and increased numbers of abortions with important consequences for the long-term viability of affected populations (Maltby and Naylor, 1990; Maltby *et al.*, 1990a, b).

Quantifying SFG relies on measuring parameters in organisms that have been exposed to pollutants and comparing these to unexposed individuals. Non-sedentary organisms can be caged in micro- and mesocosms to prevent them from migrating away from the pollutants. However, advances in microelectronics may enable some indicators of environmental stress to be measured remotely in the not-to-distant future.

8.4.4 The Energy Costs of Detoxication may Result in a Trade-off Between Production and Mortality

We have seen that pollution generally results in a loss of production (e.g. Figure 8.11). Although this can result from damage to the mechanisms of resource acquisition and uptake, it can also result from denial to production of the power and resources used by detoxication mechanisms. In the latter case, resources the organism invests in detoxication reduce its chances of death, but at a cost in terms of lost production. In other words the organism trades off a loss of production for a reduction in mortality rate.

The concept of a 'trade-off' is important in modern evolutionary ecology, and is described in detail in Chapter 13. Here we note only that the genetic possibilities for species are in general limited by trade-offs. In this section we are concerned with the trade-off between production rate and mortality rate. This can also be thought of as a trade-off between production and defence, because mechanisms which reduce mortality rate serve to defend the organism.

A trade-off between production and defence could come about in many ways (shells, spines, vigilance, etc., Sibly and Calow, 1989) but here we are especially concerned with defences against toxins. Possible methods of defence include relatively impermeable exterior membranes (e.g. Oppenoorth, 1985; Little *et al.*, 1989), more frequent moults (e.g. in Collembola, where metals stored in the gut cells are voided during ecdysis – Bengtsson *et al.*, 1985), a more comprehensive immune system, and detoxification enzymes (Terriere, 1984; Oppenoorth, 1985). Many examples are given in this book. Although it is clear such defences generally have energy costs (Sibly and Calow, 1989; Hoffman and Parsons, 1991), these could be small, for example in the case of inducible enzyme responses. Even here, however, there must be a cost, since amino acids are required at every stage of the genetic response, and since all genetic mechanisms involve overheads (molecular checking, DNA turnover, and disposal of waste). However, in environments which are naturally highly stressful (wide fluctuations in temperature, humidity etc.), the costs of coping with pollution may be only a tiny proportion of the total stress the organism has to cope with and will consequently be difficult to quantify.

To illustrate what can be achieved in this area we now consider two studies in more detail. In a classic study of insecticide detoxication in the aphid *Myzus persicae*, Devonshire and Sawicki (1979) used strains that differed in their insecticide resistance. The strains were genetically different, and the mortality and growth rates of each strain were examined in a standard environment. More resistant strains contained more detoxifying enzyme (phosphatase E_4, up to 1% of total protein) because they contained more duplications of a structural gene (Figure 8.12A). LD_{50}s were measured for three of these strains, from which mortality rates can be obtained (Figure 8.12B). Note that strains containing more copies of the structural gene, which presumably spent more on detoxication, achieved a reduction in mortality rate.

Bengtsson *et al.* (1983, 1985) studied detoxication of metals by the springtail, *Onychiurus armatus*. The springtails were fed on fungi grown on a nutrient broth contaminated with 0, 30, 90 or 300 μg g^{-1} of copper and lead in equal proportions. Concentrations of these metals within the animal reached high levels initially, but were then reduced by detoxication processes and reached steady state after a few weeks (Figure 8.13A). Detoxication was achieved by more frequent moulting of the lining of the gut (where most metals are stored), and this reduced the growth rate, as shown in Figure 8.13B. Detoxication was not complete, however, so that body metal levels were elevated in more contaminated environments even at steady state, as shown in Figure 8.13A. Mortality rate was higher in the contaminated environments (Figure 8.13B), possibly as a result of the increased metal levels in the body (Figure 8.13C). Note, however, that a certain amount of copper is needed physiologically, so

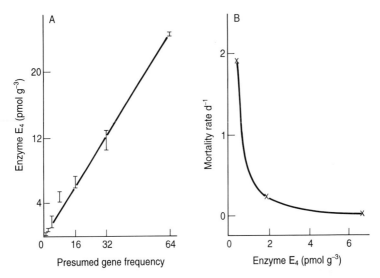

Figure 8.12 (A) Concentration of the detoxifying enzyme E_4, in seven strains of the aphid, *Myzus persicae*, in relation to the number of copies of a structural gene hypothetically present in each strain. (B) Mortality rate of three of the strains in relation to E_4 concentration. Aphids were placed on potato leaves that had been dipped in an organophosphorus insecticide (Demeton-S-methyl). Data from Sawicki and Rice (1978) and Devonshire and Sawicki (1979). From Sibly and Calow (1989).

that in a completely deficient environment growth rate must be reduced and mortality rate increased (Figure 8.13B). In this study growth rate was reduced at higher levels of pollution, presumably because the animal moulted more frequently, but metals were not completely eliminated from the body, so mortality rates were still increased.

8.5 Conclusions

Pollutants may damage organisms, directly by increasing their mortality rates, or interfering with the processes of resource acquisition and uptake and so reducing production rates. These effects on individuals result in slower population growth, or population decline (Chapter 12). Alternatively, organisms may avoid or restrict damage by the use of detoxication mechanisms or stress proteins, or repair it by e.g. DNA repair mechanisms. However, all of these use energy and resources which are therefore not available for production. It follows that the population effects are again detrimental.

Where there is a trade-off between production and defence, Figure 8.14 illustrates the likely form of the relationship. The trade-off curve has the fol-

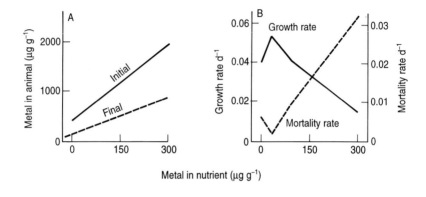

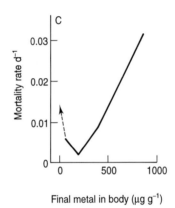

Figure 8.13 (A) Initial peak levels and final steady-state levels of metals (Cu and Pb) in the bodies of Springtails, in relation to levels in the nutrient broth. (B) Growth rate (reciprocal of time to first reproduction) and mortality rate (calculated from survivorship over the first 10 weeks of life). (C) Mortality rate in relation to the final steady-state levels of metals in the body. Mortality must increase at very low levels as the animals then suffer from copper deficiency. All data are from Bengtsson, Gunnarson and Rundgren (1983, 1985). After Sibly and Calow (1989).

lowing features: (1) even when growth rate is maximal (zero allocation to defence), growth rate is finite – equal to the limiting production rate in the study environment; (2) even if all resources are allocated to defence, mortality rate is still greater than zero, equivalent to the 'extrinsic' mortality rate in that environment. The simplest curve having these general features is convex viewed from below, and has the general form of the curve shown in Figure 8.14. The evolutionary implications of this trade-off are considered in section 13.3.

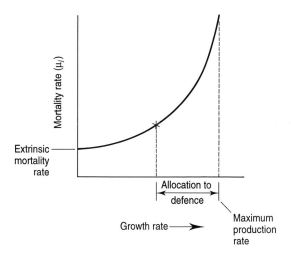

Figure 8.14   The likely form of the trade-off between production rate and mortality rate that may constrain the operation of detoxication mechanisms. Allocation of resources to defence reduces mortality rate but simultaneously cuts growth rate. After Sibly and Calow (1989).

Further Reading

HOPKIN, S. P. (1989) Includes a comprehensive review of the distribution and effects of metals in terrestrial invertebrates.

DALLINGER, R. and RAINBOW, P. S. (eds) (1993) and DONKER *et al.* (eds) (1994) Include a large number of relevant papers.

ATCHISON, G. J. *et al.* (1996) A review of the effects of pollution on the behaviour of aquatic animals.

GRUE, C. E. *et al.* (1991) Cover the effects of acetylcholinesterase on the behaviour of mammals and birds.

WIDDOWS, J. and DONKIN, P. (1992) Reviews the effects of pollution and scope for growth in *Mytilus edulis*.

9

Interactive Effects of Pollutants

In the natural environment, organisms are frequently exposed to complex mixtures of pollutants and it is relatively uncommon to find any one pollutant dominant over all others. Yet, because of limitations of time and resources, nearly all toxicity testing is carried out using single compounds. It is not feasible to test the toxicity of more than a very small proportion of the chemical combinations that exist in terrestrial, marine or freshwater ecosystems. The complexity of the situation is illustrated in Figure 9.1, which gives analytical data for residues of PCBs in tissues of organisms from a polluted area. A number of different PCB congeners are found in both species with a wider selection in the case of the mollusc, than in the harbour seal.

9.1 Introduction

When regulatory authorities consider the toxicity of mixtures, it is usually assumed (unless there is definite evidence to the contrary) that the toxicity of combinations of chemicals will be approximately additive. In other words the toxicity of a mixture of compounds will approximate to the summation of the toxicities of its individual components. This is usually a correct assumption. However, in a relatively small yet very important number of cases, toxicity is substantially greater than additive. That is to say when organisms are exposed to a combination of two or more chemicals there is *potentiation* of toxicity. Sometimes the term *synergism* is also used to describe this phenomenon. However, many scientists restrict the term 'synergism' to situations where only one of two components is present at a level that can cause a toxic effect, while the other ('synergist') would have no effect if applied alone. This practice will be followed here. Some examples of synergism are given in Table 9.1.

Table 9.1   Examples of synergism[a]

Organisms	Pesticide	Detoxifying enzyme system	Inhibitor (synergist)	Increase in toxicity
Strains of insect resistant to pyrethroids	Cypermethrin	Monooxygenase	Piperonyl butoxide	$<40\times$
Insects	Carbaryl	Monooxygenase	Piperonyl butoxide	$<200\times$
Mammals and some resistant insects	Malathion	Carboxy (B) esterase	Various organo-phophorus compounds	$<200\times$

[a] For further details see Chapter 7 of Hodgson and Levi (1993).

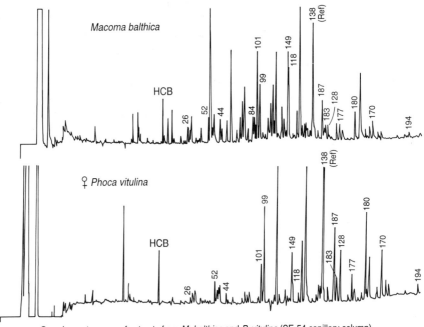

Gas chromatograms of extracts from *M. balthica* and *P. vitulina* (SE-54 capillary column). The CB congeners are identified with their IUPAC numbers.

Figure 9.1   PCB congeners in the tissues of marine organisms from the Dutch Wadden Sea. The organisms represented are a mussel (*Macoma baltica*) and the harbour seal (*Phoca vitulina*). The compounds were separated, identified and quantified by capillary gas chromatography. Each of the numbered peaks represents a PCB congener. HCB = hexachlorobenzene, used as an internal standard in the analysis. From Boon *et al.* (1989).

The effectiveness of a synergist is usually measured by a Synergistic Ratio (S.R.) which is:-

$$\frac{\text{median lethal dose or concentration for chemical alone}}{\text{median lethal dose or concentration for chemical} + \text{synergist}}$$

Where there is synergism, the SR will be greater than 1; in effect, the synergist will lower the median lethal dose or concentration of the chemical.

9.2 Potentiation of Toxicity

The phenomenon of potentiation of toxicity requires further explanation. The general picture is illustrated in Figure 9.2, where two compounds A and B are under consideration. The maximum dose of either of them will give the same degree of toxic response (X). (X could be a percentage mortality). Between the two maximum doses are different mixtures of the two compounds. Doses for either compound run from 0 to 100% of the maximum dose. The summation of the contributions of the two components of the mixture will always be 100%, e.g. 30% of maximal dose of A + 70% of the maximal dose of B. If toxicities are simply additive, all of these combinations should give the same toxic response (X) as the maximal dose of A or B. If, however, there is potentiation or synergism, the toxic effect should be greater than expected. If there is antagonism, the toxic effect should be less.

Care is needed when deciding whether toxic effects of combinations of chemicals are truly greater than additive. In the first place, due to errors in measurement, interest is confined to situations where toxicity is *substantially* greater than additive (e.g. where SRs exceed 2). Smaller differences usually

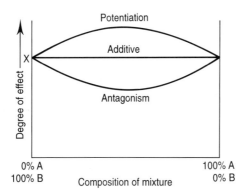

Figure 9.2 Potentiation of toxicity. The vertical axis indicates the quantity of the compound, and the horizontal axis represents time. The maximum dose of compounds A and B both give the same degree of toxic response X. Potentiation is seen when the toxicity of a mixture of two compounds exceeds the summation of toxicities of the individual components. (After Moriarty, 1988.)

reflect no more than the compounding of errors. Also, account should be taken of the relationship between dose and toxic effect for each of the individual components of a mixture (Figure 9.3). In particular, it is important to know whether there is a straight-line relationship between dose (or log dose) and toxic response (Figure 9.3). If this is the case, then increases in toxicity of combinations of chemicals which are substantially greater than additive are to be regarded as examples of potentiation. If, on the other hand, they are not linear – e.g. where the increase in toxicity of an individual compound is proportionately greater than the corresponding increase in dose – this conclusion does not necessarily follow (Figure 9.3). An enhancement of toxicity above that which is simply additive may merely reflect what happens when the dose of an individual chemical (or chemicals) is increased, and may not therefore represent potentiation due to interaction between chemicals. Having said this, it should be re-emphasised that this chapter is mainly concerned with cases where potentiation represents considerable enhancement of toxicity – frequently of an order of magnitude (i.e. 10-fold or more).

The identification of combinations of pollutants which give rise to problems of potentiation might seem an impossible task. However, there are guidelines

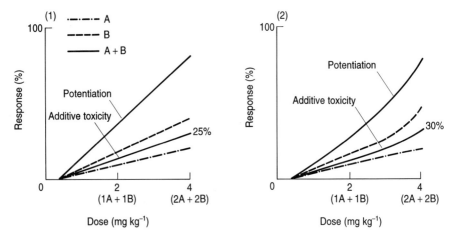

Figure 9.3 Additive toxicity and potentiation. In (1) both compounds A and B produce a linear response over the dose range 0–4. If toxicity is simply additive, the response to 1 mg kg^{-1} A + 1 mg kg^{-1} B is intermediate between the responses to 2 mg kg^{-1} A and 2 mg kg^{-1} B. If potentiation occurs then the response to the combination will be higher than is shown for the additive response.

In (2) the response to A is linear, but the response to B is non-linear. In this situation the additive response to 2 mg kg^{-1} A + 2 mg kg^{-1} B is greater than in (1). This illustrates the point that linearity cannot be *assumed* for individual response curves. If, in this case, dose–response curves were known up to 2 mg kg^{-1} for A and B, and the response to 2 mg kg^{-1} A + 2 mg kg^{-1} B were as shown – it might be wrongly assumed that potentiation had occurred. To establish that potentiation has occurred, the dose–response curves for the compounds A and B need to be determined above the doses used in combination.

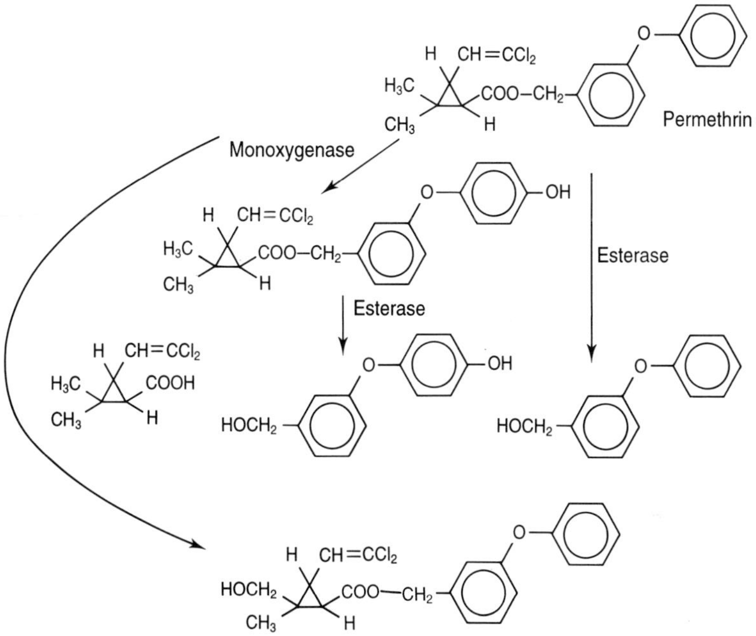

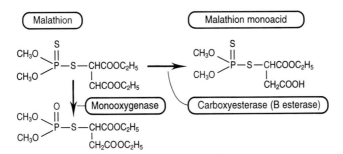

Figure 9.4 Metabolism of permethrin. Permethrin is detoxified by two different systems – monooxygenase which attacks both acid and the alcohol moieties, and esterase ('B' esterase) which breaks the carboxyester bond. Inhibitors of monooxygenase can increase the toxicity of this and other pyrethroids.

Figure 9.5 Metabolism of malathion. Malathion is detoxified by the action of a carboxy esterase but activated by monooxygenase. Toxicity depends on the relative importance of these two competing enzymes.

Chemicals which induce the monooxygenase system can make malathion more toxic.

169

which aid the recognition of such combinations. In particular, recent rapid advances in biochemical toxicology have given more insight into the potentiation of toxicity due to interactions at the toxicokinetic level (Chapter 5). When one compound (A) causes a change in the metabolism of another (B), two types of interaction are recognized:

1. Compound A inhibits an enzyme system that detoxifies compound B. Thus, the rate of detoxication of B is slowed down because of the action of A.
2. Compound A induces an enzyme system which activates compound B. Thus, the rate of activation of B is speeded up due to the action of A.

These two phenomena will now be considered separately.

9.3 Potentiation due to Inhibition of Detoxication

In terrestrial vertebrates and invertebrates, the effective elimination of lipophilic xenobiotics depends upon their enzymic conversion to water-soluble products which are readily excreted (Chapter 5).

The inhibition of enzymes concerned with detoxication can lead to an increase in the toxicity of the compounds that they metabolize. This general phenomenon is well illustrated by the effect of synergists upon insecticide toxicity (Table 9.1). Some of the most striking examples of synergism involve inhibition of monooxygenases by piperonyl butoxide and other methylenedioxyphenyl compounds. Synergists of this type can increase the toxicity of pyrethroid and carbamate insecticides by as much as 60-fold and 200-fold, respectively. (The synergistic ratios provide measures of increases in toxicity.)

Box 9.1 Two examples of interest in ecotoxicology

Pyrethroid insecticides, e.g. cyhalothrin, have considerable toxicity to bees. Often, however, they do not cause much damage when they are properly used in the field. Bees are repelled by pyrethroids, perhaps as the consequence of low-level sub-lethal effects. However, pyrethroids can become very much more toxic in the presence of certain 'EBI' fungicides (EBI = ergosterol biosynthesis inhibitor). Synergistic ratios of 5–20 have been reported for bees when EBI fungicides are added to pyrethroid insecticides. This potentiation of toxicity has been attributed to the inhibition of detoxication of pyrethroids by the monooxygenase system (Figure 9.4).

In a second example, organophosphorus insecticides which contain the thion (P = S) group, can inhibit microsomal monooxygenases of vertebrates. This happens when the enzyme converts them to oxons (P = O) (Figure 9.5). The process is termed 'oxidative desulphuration' and leads to the binding of sulphur to cytochrome P_{450} with consequent loss of monooxygenase activity. Exposure to relatively low levels of organophosphorus compounds can make birds more sensitive to the toxicity of carbamate insecticides; carbamates are detoxified by MOs, so a reduction of MO activity can cause an increase in toxicity (Figure 9.5).

9.4 Potentiation due to Increased Activation

Metabolism of lipophilic xenobiotics brings a reduction of toxicity in the great majority of cases. However, there are some very important exceptions to this rule. Oxidation by the monooxygenase systems sometimes generates highly reactive metabolites which can cause cellular damage. In principle, therefore, the induction of MO by a non-toxic dose of one compound can increase the toxicity of a second compound which is subject to oxidative activation. This said, it does not follow that the induction of monooxygenase will automatically lead to increases in the rate of activation, (or in the degree of consequent cellular damage). For one thing, induction may also lead to the induction of other enzymes which have a detoxifying function and can compensate for increased activation.

Box 9.2 Two examples where potentiation results from increased activation

Benzo(a)pyrene and certain other carcinogenic polycyclic aromatic hydrocarbons (PAHs) are activated by an inducible form of cytochrome P_{450} known as P_{450} 1A1 (Chapter 5). A range of planar organic compounds, not in themselves carcinogens or mutagens, can cause the induction of P_{450} 1A1. Examples include certain PAHs, coplanar PCBs and 1,2,7,8-tetrachlorodibenzodioxin (TCDD) (see Chapter 7). Such compounds can act as promoters, which potentiate the carcinogenic action of other compounds. By increasing the rate of activation of carcinogens, they can also increase the rate of formation of DNA adducts; this may lead to an increased rate of chemically-induced mutation. In the marine environment there is much evidence that (1) fish, birds and mammals sometimes have elevated levels of P_{450} 1A1, and (2) that the levels of P_{450} 1A1 are related to the degree of exposure to pollutants such as coplanar PCBs. It is suspected, but not yet proven, that individuals with elevated 1A1 will experience higher levels of DNA damage caused by environmental carcinogens and mutagens.

In a second example, OP insecticides, which contain P = S group, are activated by P_{450} forms of the monooxygenase system (Figure 9.4). Thus, the induction of mono-oxygenase by a non-toxic dose of a xenobiotic can lead to enhanced activation and consequently to increased toxicity of an OP. Partridges exposed to EBI fungicides can show increased levels of hepatic monooxygenase due to induction. While in the induced state they show increased susceptibility to OP insecticides such as malathion and dimethoate (Walker *et al.*, 1993; Johnston, 1995). The time dependence of this potentiation needs to be emphasized. Following exposure to the EBI, several hours will elapse before the rate of activation of the OP is increased. Furthermore, the mono-oxygenase activity will return to its normal level after a few days, if exposure to the EBI does not occur again. Thus birds may become more susceptible to certain OPs during the period 6 h–5 days following exposure to an EBI. Potentiation of this kind has, so far, been demonstrated only in laboratory studies. It has not yet been shown to occur in the field with normal approved use of pesticides (but see further discussion in next section).

9.5 The Detection of Potentiation in the Field

While potentiation is a well recognized phenomenon in the laboratory, the extent to which it occurs in the field is virtually unknown. There are good reasons for suspecting that it may occur in heavily polluted areas. In some marine areas, for example, a wide range of organochlorine-compounds (PCBs, TCDDs and others) (Figure 9.1) are found (Malins and Collier, 1981). Elevated levels of P_{450} 1A1 have been found in fish and birds from such areas and relatively high rates of activation of mutagenic PAHs have been suspected.

In intensive agriculture and horticulture, animals, birds and non-target invertebrates are exposed to a variety of insecticides, herbicides and fungicides. In particular, combinations of pesticides are sometimes used, as seed dressings and as sprays (e.g. tank mixes). Questions are now asked about the possibility of potentiation of toxicity when animals are exposed to mixtures such as these. Also, mobile species (e.g. birds, flying insects) can be exposed sequentially to different compounds when they move from field to field. As explained earlier, there is good evidence that bees can be poisoned by pyrethroid sprays when EBIs are mixed with them; EBIs can cause potentiation by the inhibition of detoxication (Pilling *et al.*, 1995). There are also other combinations of pesticides which give cause for concern in regard to possible potentiation of toxicity under field conditions. However, these issues remain the subject of speculation, because field studies which could resolve them have yet to be performed.

The difficulty of identifying harmful effects caused by pollutants in the field is a recurring theme of this book. The potential of biomarker strategies to aid the resolution of this problem was emphasized in the previous chapter. This point can be illustrated by considering a hypothetical example, where the interaction of two pesticides is investigated in a field trial. One or more biomarkers of toxic effect would be chosen to measure responses to a pollutant in an appropriate indicator species. These responses would then be determined in three different field situations:

1. where only compound A was applied;
2. where only compound B was applied;
3. where both A and B were applied.

If potentiation occurred then the responses measured in 3 would be substantially greater than the summation of responses measured in 1 and 2. This approach has the advantage that it can measure potentiation at the sublethal level, i.e. it can give early warning of enhancement of toxicity before lethal effects are produced.

There are advantages in using non-destructive biomarkers in situations of this kind. In particular they make it possible to conduct serial sampling in individual animals or birds; any changes caused by chemicals can then be measured in relation to *internal* controls. In other words biochemical or physiological parameters can be measured in individuals before and after exposure to the pesticide, thereby overcoming the serious difficulty of inter-individual variation. Serial sampling is possible, for example, in the case of nestling birds in nest boxes, or in animals or birds which are restricted to limited areas (e.g. enclosures) from which they can readily be recaptured and sampled.

Further Reading

MALINS, D. C. and COLLIER, T. K. (1981) Discusses the problem of effects of mixtures upon marine organisms.

BOON, J. P. *et al.* (1989) Gives examples of complex pollution patterns caused by PCBs.

WALKER, C. H. *et al.* (1993) Discusses laboratory evidence for potentiation in birds.

WILKINSON, C. F. (1976) Deals with potentiation of toxicity of insecticides.

MORIARTY, F. (1988) *Ecotoxicology*, 2nd Edition Discusses the problem of interpretation of data purporting to show potentiation.

10

Biomarkers

The term 'biomarker' has been gaining acceptance in recent years, albeit with some inconsistency in definition. Here we define biomarkers as a 'biological response to a chemical or chemicals that gives a measure of exposure and sometimes, also, of toxic effect'.

Biological response is, of course, a very broad term. In theory it could apply to anything from binding to a receptor to the functioning of an ecosystem. In practice, biomarkers are generally considered to run from binding of receptors to effects on whole animals (see Introduction). Despite the importance of any changes at the population or community structure level (considered in more detail in Chapters 12 and 15), these changes are too general to be considered as specific biomarkers. Some examples of biomarkers at various organizational levels are given in Table 10.1.

Table 10.1 Biomarkers at different organizational levels

Organizational level	Example of biomarker
Binding to a receptor	TCDD binding to Ah receptor Nonylphenols binding to oestrogen receptor
Biochemical response	Induction of monooxygenases Vitellogenin formation
Physiological alterations	Eggshell thinning Feminization of embryos
Effect on individual	Behavioural changes Scope for growth

10.1 Classification of Biomarkers

A number of classifications of biomarkers have been proposed. The most widely used is division into biomarkers of exposure and biomarkers of effect. Biomarkers of exposure are those that indicate exposure of the organism to chemicals, but do not give information of the degree of adverse effect that this change causes. Biomarkers of 'effect', or more correctly 'toxic effect' (since all biomarkers by definition show an effect), are those which demonstrate an adverse effect on the organism.

A biomarker approach based on changes in physiological parameters is shown in Figure 10.1. The change in the health status of an individual with increasing exposure to a chemical is shown by a smooth curve running from

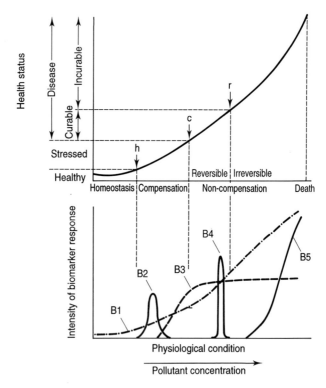

Figure 10.1 Relationship between exposure to pollutant, health status and biomarker responses. Upper curve shows the progression of the health status of an individual as exposure to pollutant increases. h = the point at which departure from the normal homeostatic response range is initiated; c = the limit at which compensatory responses can prevent development of overt disease; r = the limit beyond which the pathological damage is irreversible by repair mechanisms. The lower graph shows the response of five different hypothetical biomarkers used to assess the health of the individual. From Depledge *et al.* (1993).

Table 10.2 Some biomarkers listed in order of decreasing specificity to pollutants

Biomarker	Pollutant	Comments and references to analytical procedures
Inhibition of ALAD	Lead	Sufficiently reliable to replace chemical analysis (Wigfield et al., 1986)
Induction of metallothionein	Cadmium	More difficult to measure than cadmium levels (Hamer, 1986)
Eggshell thinning	DDT, DDE, Dicofol	Degree of eggshell thinning is easily measured (Ratcliffe, 1967)
Inhibition of AChE	OPs, Carbamates	Easier and more reliable than chemical analysis (Fairbrother et al., 1991)
Anticoagulant clotting proteins	Rodenticides	Measurements comparable in complexity to chemical analysis (Huckle et al., 1989)
Induction of monooxygenases	OCs, PAHs	Dioxin equivalent more easily measured than using chemical analysis (Safe, 1990)
Porphyrin profiles	Several OCs	Separation by high-pressure liquid chromatography is well developed (Kennedy and James, 1993)
Retinol profiles	OCs	Considerable natural variations, ratios more reliable than absolute values (Spear et al., 1986). Provides means of showing exposure to specific chemicals (Shugart, 1994)
DNA and haemoglobin adducts	Largely PAHs	Several tests available but complicated by repair mechanisms
Other serum enzymes	Metals, OCs, PAHs	Several different enzyme systems have been studied (Fairbrother, 1994)
Stress proteins	Metals, OCs	Wide range of stress proteins have been studied (Sanders, 1993)
Immune responses	Metals, OCs, PAHs	Many different tests are available (Wong et al., 1992)

healthy through reversible to irreversible changes leading to death. The important transition points along the way are (1) when the organism is first stressed (h), (that is when physiology is no longer normal), then moving into the area where the organism, although stressed, is able to compensate for this stress, (2) when the organism is no longer able to compensate (point c), but the changes are still reversible and removal of the stress enables the organism to recover, and (3) point r beyond which the changes are irreversible and death ensues. The second part of Figure 10.1 shows the responses of five different biomarkers which are used to measure the health status of the individual.

10.2 Specificity of Biomarkers

Biomarkers range from those that are highly specific – an enzyme of the haem pathway aminolevulinic acid dehydratase (ALAD) is inhibited only by lead – to those that are non-specific, whereas effects on the immune system can be caused by a wide variety of pollutants. A listing of biomarkers in order of their degree of specificity is given in Table 10.2.

Both highly specific and highly non-specific biomarkers are of value in hazard assessment. When blood samples of waterfowl can be taken and the activity of ALAD determined then it is possible, without further measurements, to determine the percentage of the waterfowl that are at risk from lead poisoning. However, determination of ALAD does not give any information on the (many) other pollutants that may be present.

The inhibition of AChE is specific to the organophosphorus and carbamate pesticides and inhibition of this enzyme in the brain would be legally accepted as proof of death by one of these agents (Hill and Fleming, 1982). It is more easily and reliably measured than the residues of the pesticides themselves which are often difficult to determine chemically, and many of them break down rapidly in the body. Nevertheless, to prove which of these agents was responsible, chemical analysis would be necessary. So the inhibition of AChE is a valuable biomarker for investigations on agricultural land or areas likely to be affected by run-off from these areas, but would not be used in investigations outside these areas.

The induction of monooxygenases is caused by a wide variety of chemicals. Thus it is a useful indication that organisms are affected by pollutants although it gives little information on the specific cause. In this case the biomarker is valuable to target additional studies, since pollutants are present at a high enough level to cause a detectable effect.

10.3 Relationship of Biomarkers to Adverse Effects

It is clearly useful to be able to relate the degree of change of the biological response measured to the harm that it causes so that when remedial action is proposed the cost of it can be defended. A list of the same biomarkers already

Table 10.3 Some biomarkers listed in order of decreasing specificity of adverse effect

Biomarker	Organizational level	Comments and references
Eggshell thinning	Intact animal – population	Wide species variation in sensitivity. Related to reproductive success (Peakall, 1993)
Inhibition of AChE	Organ – intact animal	Degree of inhibition has been related to mortality and sub-lethal effects (Grue et al., 1991)
Inhibition of ALAD	Organ – intact animal	Degree of inhibition has been related to mortality (Scheuhammer, 1989)
Anticoagulant clotting proteins	Intact animal – population	Has been related to mortality, risk assessed from blood protein levels (Hegdal and Blaskiewicz, 1984)
Depression of plasma retinol and thyroxine	Organ	Binding to specific protein has been shown. Relation to adverse effects tenuous (Brouwer and van den Berg, 1986)
Induction of monooxygenases	Organ population	Analysis of dioxin equivalents has been related to reproductive success. Induction of P_{450} enzymes related to specific chemicals (Bosveld and van den Berg, 1994)
DNA integrity	Organ	DNA damage is a serious indication of harm, but relationship to effects often tenuous (Everaarts et al., 1993)
Immune responses	Organ	Proper functioning is critical to health, but system has considerable reserve (Richter et al., 1994)
DNA and haemoglobin adducts	Organ	Good monitor of exposure, especially for PAHs, relation to effects is tenuous (Varanasi et al., 1989)
Other enzymes	Organ	Relationship to effects are not clear (Fairbrother, 1994)
Porphyrin profiles	Organ	Levels in environmental samples are well below those causing adverse effects (Fox et al, 1988)
Induction of metallothionein	Organ	Protective mechanism, not related to mechanism of toxicity (Hamer, 1986)
Stress proteins chemicals	Organ	Difficult to separate effects from non-chemical stresses (Sanders, 1993)

given in Table 10.2 are listed in Table 10.3 in the order of our knowledge of the adverse effects that they measure. A quick examination of the two tables will reveal that they differ in ranking order.

At the top of Table 10.3 is eggshell thinning. In this case it is possible to define the critical degree of eggshell thinning: it has been found for a variety of species that eggshell thinning in excess of 16–18% is associated with population declines. This phenomenon is discussed in more detail in Chapter 15.

The fact that the relationship between the biomarker response and an adverse effect is not clear-cut does not invalidate the use of that biomarker. First, it demonstrates that the organism has been sufficiently exposed to a pollutant or pollutants to cause a physiological change. In some cases, such as the induction of metallothionein, the change is a protective mechanism (Chapter 7); here a knowledge of how much of the possible protective mechanism has already been induced is valuable to assess the risk to the individual. Second, in the case of vital systems, it is an indication that further investigations should be undertaken. For example, few would take damage to the integrity of DNA lightly even though in many cases the damage is repaired and no adverse effects occur (see Chapter 7). In other cases, such as changes in porphyrin levels, it is clear that the levels are much lower than has been shown to cause harm. Nevertheless these biomarkers can be used to demonstrate exposure. The use of these various types of biomarkers in hazard assessment is considered later in this chapter.

10.4   Discussion of Specific Biomarkers

Some of the biomarkers listed in Tables 10.2 and 10.3 have been covered elsewhere in this book. Specifically, eggshell thinning is discussed in Chapter 15, and some information on the inhibition of AChE and induction of monooxygenases has been given in Chapter 7. The sections that follow discuss some, but by no means all, of the available biomarkers. More complete coverage is given in the books on biomarkers given in the reading list at the end of this chapter.

10.4.1   *Inhibition of Esterases*

From the point of view of ecotoxicology, AChE is particularly useful as it represents the site of action, and its degree of inhibition is related to toxic effects. Butyrylcholinesterase (BuChE) is sometimes studied in parallel with AChE but its physiological role is unknown, and its degree of inhibition is not simply related to toxic effect. The study of neuropathy target esterase the interaction of which with organophosphorus compounds (OPs) can lead to organophosphorus-compound-induced delayed neurotoxicity has been confined to laboratory studies.

The mode of action is well established and has already been considered in some detail in Chapter 7. Two classes of compounds, the OPs and carbamates, inhibit AChE causing an accumulation of acetylcholine at the nerve synapses and disruption of nerve function. Disruption of nerve function has obvious effects: tremors, motor dysfunction and death. The assay for AChE is more straightforward, quicker and cheaper than chemical analysis for OPs or carbamates. The degree of inhibition of AChE has been related to the symptoms observed.

With vertebrates, the inhibition of brain AChE has often been used to establish that death has been caused by OP or carbamate pesticides. Under ideal conditions, inhibition in the range of 50–80% can be taken as proof of mortality from the pesticide (Hill and Fleming, 1982). In practice, the degree of denaturation is often unknown and adequate controls are often difficult to obtain. Also, inhibition by carbamates is readily reversible (cf OPs) and can be quickly lost after death. However, the measurements of OP and carbamates themselves are even more difficult since they are rapidly metabolized and eliminated; in fact chemical analysis for residue levels has not been widely used in the diagnosis of poisoning by these compounds. The use of AChE inhibition is discussed in the diagnosis of damage caused by pesticides in Chapter 15 as a consequence of forest spraying in Eastern Canada.

The usual organ studied in the case of wild vertebrate samples is the brain, which is the principal site of action of OPs and carbamates. While using inhibition of blood AChE or BuChE would be more acceptable, the relationship to inhibition of brain AChE is complex. Studies have shown that variability of esterase activity is much greater with plasma than with brain and that recovery of plasma AChE activity is much more rapid than in the case of brain AChE. Also, plasma AChE does not represent the site of action of OPs and carbamates and there is no simple relationship between degree of inhibition and toxic effect. Thus diagnosis based on plasma AChE activity is difficult.

10.4.2 *The Monooxygenases*

The haem-containing enzymes known as cytochromes P_{450} are major components of the defences of organisms against toxic chemicals in their environment. Originally evolved, perhaps as long as two thousand million years ago, to handle naturally occurring toxic compounds (Nebert and Gonzalez, 1987), they now play an important role in the detoxication of man-made chemicals.

The monooxygenase system is a coupled electron-transport system composed of two enzymes – a cytochrome and a flavoprotein (NADPH-cytochrome reductase). The system occurs in the endoplasmic reticulum of most organs, but the activity is much higher in the liver than most other tissues (Chapter 5). Initially these two enzymes were divided into two major classes, the enzymes P_{450} and the enzymes P_{448} (the numbers refer to the wavelengths at which the cytochrome could be detected). Recent work has shown the complexity of the system. A recent listing gave 78 members divided into 14 families. Under this system, P_{448} becomes $P_{450}I$, and the enzymes P_{450} are represented in the 13 remaining families. Cytochrome $P_{450}I$-dependent reactions include N-oxidation and S-oxidation; widely studied enzyme activities include ethoxyresorufin O-deethylase (EROD), benzo(a)pyrene hydroxylase, (BaPH) and aryl hydrocarbon hydroxylase (AHH). Cytochrome $P_{450}II$-dependent oxidations include aromatic hydroxylation, acyclic hydroxylation, dealkylation and deamination. The most widely studied enzyme activities are benzphetamine N-demethylase and aldrin epoxidase.

The first demonstration of the effect of an organochlorine pesticide on hepatic microsomal metabolism (see Chapter 5) was made over 30 years ago. This important finding was the fortuitous result of an experiment to examine the effects of food deprivation on drug metabolism. These workers noted an unexpected decrease in sleeping time of rats given phenobarbital after their cages were sprayed with chlordane. They discovered that this was due to stimulation of hepatic microsomal drug metabolizing enzymes (Hart *et al.*, 1963), and this finding was soon extended to many other organochlorines. This includes not only all the organochlorine pesticides, but also the PCBs, PCDFs and PCDDs. The assessment of complex families of chemicals by means of the calculation of dioxin equivalents is considered in Chapter 15.

Activity of monooxygenases is affected by a wide variety of compounds. Classes of compounds of environmental interest besides the organochlorines include the organophosphorus compounds, pyrethroids and polycyclic aromatic hydrocarbons (PAHs).

The concept of using induction of monooxygenase activity in fish as a monitor of pollution of the marine environment by oil was put forward by several workers in the mid 1970s. Since then a wide variety of studies have been published ranging from the induction of AHH in fish off the Los Angeles sewage outlets to EROD induction by pulp mill effluent in Sweden. Induction of monooxygenases by paper mill effluent has proved to one of the most sensitive biomarkers for tracking this particular type of pollution.

Monooxygenase activity is shown by a wide range of species (see Chapter 5 and 7) and some studies on fish-eating birds in the Great Lakes are detailed in Chapter 15. The usefulness of monooxygenases for biological monitoring has been clearly demonstrated in the case of hydrocarbon pollution in fish and aquatic invertebrates and for both PAHs and OC contamination in a wide range of organisms. Since the response is caused by a very wide variety of chemicals, it means that the system is capable of detecting exposures sufficiently high to cause a biological response to many xenobiotics. Conversely it tells little about the causative agent(s), but can be used to delimit an area for which it is worth the time and expense of more detailed investigations.

From a practical viewpoint, the considerable variation within a specific population means that the sample size usually has to be fairly large. The fact that the system is induced by a large variety of natural compounds as well as xenobiotics and is affected by a wide variety of other parameters – temperature, diet etc. – means that great care must be taken to ensure that there are reliable control levels.

10.4.3   *Studies on Genetic Material*

The fundamental role of DNA in the reproductive process is well known and will not be discussed further here. The endpoints used to assess the damage caused to DNA adducts by environmental pollutants are specific genotoxic effects, especially the increase of carcinogenesis, rather than effects on the reproductive process.

There is a sequence of events between the first interaction of a xenobiotic with DNA and consequent mutation, which may be divided into four broad categories. The first stage is the formation of adducts. At the next stage, there may be secondary modifications of DNA such as strand breakage or an increase in the rate of DNA repair. The third stage is reached when the structural perturbations to the DNA become fixed. At this stage, affected cells often show altered function. One of the most widely used assays to measure chromosomal aberrations is sister chromatid exchange. Finally, when cells divide, damage caused by toxic chemicals can lead to the creation of mutant DNA and consequent alterations in gene function.

The covalent binding of reactive metabolites of environmental pollutants to DNA – adduct formation – is a clear demonstration of exposure to these agents and an indication of possible adverse effects. The relationships between environmental levels, the degree of adduct formation and the ultimate effect are complex. For example, although a direct relationship between the extent of cigarette smoking and the number of DNA–BaP adducts has been clearly shown, the relationship between DNA–BaP adducts and the occurrence of lung cancer is less well defined. In the field of wildlife toxicology the establishment of the sequence of events from the initial DNA lesion to harm is even more difficult. Nevertheless, it is reasonable to conclude that reaction of chemicals with DNA can have harmful consequences such as tumour formation.

Two different approaches have been used to study the formation of DNA adducts after exposure to pollutants:

1. radioactive post labelling (usually with ^{32}P) leading to the separation of a range of adducts by two-dimensional thin-layer chromatography;
2. techniques to identify specific adducts, including fluorescence spectrometry, chromatographic techniques and enzyme-linked immunosorbent assays (ELISA) – techniques which are sensitive, if properly employed, and can detect one adduct among 10^8 normal nucleotides.

The two techniques give different information. The former gives an index of the degree of total covalent binding, whereas the latter gives information on the actual degree of binding for a few specific compounds.

Monitoring of adduct formation provides one of the best means to detect exposure to polycyclic aromatic hydrocarbons (PAHs). The stability of DNA and haemoglobin adducts formed by this class of compounds means that evidence of exposure to them remains after they have been cleared from the body. The fact that adduct formation by haemoglobin can be studied means that non-destructive testing is possible. The exact relationship between adduct formation and carcinogenesis is under intensive study related to human health. In the environmental field the correlation between macromolecular damage and the epidemiological data is in an earlier stage.

In studies at Puget Sound, Washington, fish and sediments were sampled. The levels of PAHs in sediment and gut were determined and the extent of DNA–xenobiotic adducts in the liver measured by ^{32}P labelling. Additionally, concentrations of PCBs and the degree of induction of MFOs were determined. The various indices enabled these workers to discriminate between sites which exhibited a considerable range of differences in chemical contamination by both PCBs and PAHs (Stein *et al.*, 1992).

Breakage in chromosomes can be examined directly under the microscope or by the alkaline unwinding assay (Peakall, 1992). The latter technique is based on the fact that DNA strand separation takes place where there are breaks. The amount of double-stranded DNA remaining after alkaline unwinding is inversely proportional to the number of strand breaks, provided that renaturation of the DNA is prevented.

Chromosome breaks in the gills of the mud minnow have been used to study pollution of the Rhine. Also, increased chromosomal aberrations were found in rodents collected from areas close to a petrochemical waste disposal site. Although damage to chromosomes can lead to serious effects it must be remembered that repair mechanisms are capable of preventing this from happening.

Sister chromatid exchange (SCE) is the reciprocal interchange of DNA during the replication of chromosomal DNA. Chromosomes of cells that have gone through one DNA replication in the presence of labelled thymidine or nucleic acid analogue 5-bromodeoxyuridine and then replicated again in the absence of the label are generally labelled in only one of the chromatids. The label is observed to be exchanged from one chromatid to the other. The sister chromatid exchanges can be easily visualized using the light microscope or by differential staining.

Chromosomes that have undergone SCE should not be regarded as damaged in the conventional sense since they are morphologically intact. Nevertheless, SCE occurs at sites of mutational events including chromatid breakage. Good correlations have been observed between the number of induced SCEs per cell against the dosage of X-ray and the concentration of a number of chemicals known to cause chromosomal aberrations.

A relationship between SCE level (Nayak and Petras, 1985), and distance from an industrial complex, was demonstrated in wild mice in Ontario, Canada, and a variety of chromosomal aberrations were found in cotton rats living close to hazardous waste dumps in the United States. Fish exposed to water from the Rhine showed a marked increase of SCE levels (van der Gaag *et al.*, 1993).

As has already been mentioned under specific assays a number of changes in the genetic material can be used to monitor for pollution. Monitoring has been carried out on the incidence of tumours in fish. Fish are frequently sampled in considerable numbers and, more importantly, many of their tumours are visible externally. Such data could not be readily collected from other classes of species even when large sample sizes are available such as muskrat from trappers, or duck from hunters, due to the cost of dissection.

In the North American Great Lakes surveys to determine the instance of tumours have been carried out as part of the surveillance programme. The levels of occurrence of tumours in brown bullheads (*Ictalurus nebulosis*) and white suckers (*Catostomus commersoni*) are greatest in the most polluted areas. Given the large number of contaminants in the Great Lakes it is virtually impossible to link carcinogenesis to a specific chemical, but circumstantial evidence for a chemical origin is strong in many cases, although it should be cautioned that viral agents and parasites can cause neoplasmas. The high incidence of tumours in fish in rivers highly contaminated with PAHs strongly suggests that these agents are involved (Environment Canada, 1991).

Overall, the studies involving DNA have reached an interesting stage. A great deal of medical research is being carried out and there are strong indications that this information can be used in assessing the impact of pollutants, especially PAHs, on wildlife.

10.4.4 *Porphyrins and Haeme Synthesis*

Porphyrins are produced by the haeme biosynthetic pathway which is a vital system for most of the animal kingdom. Two major disruptions of haeme biosynthesis by environmentally important agents have been studied. These are the formation of excess amounts of porphyrins by some organochlorines (OCs) and the inhibition by lead on the enzyme aminolevulinic acid dehydratase (ALAD).

Haeme biosynthesis is normally closely regulated, and levels of porphyrins are ordinarily very low. Hepatic porphyria is characterized by massive liver accumulation and urinary excretion of uroporphyrin and heptacarboxylic acid porphyrin. While the mechanism of OC-induced porphyria has not been completely elucidated, it is considered by several workers that inhibition of the enzyme uroporphyrinogen decarboxylase is the proximal cause. The two OCs that are most involved in inducing porphyria are hexachlorobenzene (HCB) and the PCBs. Although HCB has been shown, in both mammals and birds, to induce porphyria, the dosages required are high compared to environmental levels. PCBs have also been shown to be potent inducers, although the various congeners act quite differently.

Studies on the Rhine showed that the patterns of hepatic porphyrins were markedly different and the total porphyrin levels much higher in pike collected than those from the cleaner River Lahn. The levels of organochlorines were up to 40-fold higher in the fish from the Rhine than those from the Lahn.

The levels of hepatic highly carboxylated porphyrins (HCPs) were markedly elevated in herring gulls (*Larus argentatus*) collected from the North American Great Lakes when compared to those from the Atlantic coast (Fox *et al.*, 1988).

The variation in the mean of the hepatic levels of HCPs in seven species of birds (covering five orders) was only two-fold and the total range was 4–22 pmol g^{-1} (Fox *et al.*, 1988). No comparable study appears to have been made on any other class of organism. These baseline data, collected from areas of low contamination, show only small variation, but in view of variability of response to OCs in experimental studies, variability in areas of high contamination is a problem.

Aminolevulinic acid dehydratase (ALAD) is an enzyme in the haeme biosynthetic pathway. Inhibition of ALAD was first studied over 25 years ago as a means of detecting environmental lead exposure in humans and has since become the standard bioassay for this purpose; it has subsequently been used in wildlife investigations. The assay is highly specific for lead, with other metals being 10 000 times less active in causing inhibition. ALAD inhibition is rapid. The effect is only slowly reversed, with ALAD values returning to normal values only after about 4 months.

Inhibition of ALAD has been used as an indicator of lead exposure both for general problems, such as in urban areas and along highways, and also specifically to study the 'lead-shot problem' in waterfowl. A three-fold difference in the blood ALAD activity was found between rats in a rural and an urban site in Michigan (Mouw *et al.*, 1975). The main physiological indications of lead toxicity in the urban rats were an increase in kidney weight and the incidence of intranuclear inclusions. Both effects could be correlated with lead levels. Similarly, marked differences in the ALAD activity were found between feral pigeons (*Columbia livia*) from rural, outer urban, suburban and central London areas (Hutton, 1980).

The lead levels, ALAD activity and reproductive success of barn swallows (*Hirundo rustica*) and starlings (*Sturnus vulgaris*) along highways with different traffic densities have been studied in North America. It was found that there was a significant increase of the lead levels in the feathers and carcasses of both adults and nestlings and 30–34% decrease in plasma ALAD activity. However, the number of eggs laid, number of young fledged, and pre-fledgling body weights were not affected, indicating that lead from automotive emissions does not pose a serious hazards to birds nesting close to motorways.

The inhibition of ALAD has been shown to be a reliable indicator of exposure to lead in studies on several species of fish. A linear regression was found when ALAD activity was plotted against both concentration of lead in the blood and in the water.

Mortality of ducks and other waterfowl due to the ingestion of lead shot has been of serious concern for many years, the issue first being raised in North America over 70 years ago. The problem is caused by ducks and geese ingesting spent lead shot during the course of their feeding. A nation-wide survey found that 12% of the gizzard samples examined contained at least one lead shot and considered that 2–3% of all waterfowl in North America died from lead poisoning. Secondary poisoning of bald eagles feeding on waterfowl has also been of concern. National surveys of eagles found dead in the United States showed that about 5% died from lead poisoning. The inhibition of ALAD has been shown to be sensitive enough to detect the effect of a single pellet. A strong negative correlation between blood lead concentration and log ALAD activity has been found by many workers. The ALAD assay is a simple one, which can be carried out without expensive equipment or lengthy training.

ALAD inhibition represents one end of the biomarker spectrum. It is a sensitive, dose-dependent measurement that is specific for a single environmental pollutant, lead.

10.4.5 Behavioural Biomarkers

Behavioural changes represent a higher organizational level of biomarker than any considered so far. One of the early proponents of the value of behavioural toxicology stated that 'the behaviour of an organism represents the final integrated result of a diversity of biochemical and physiological processes. Thus, a single behavioural parameter is generally more comprehensive than a physiological or biochemical parameter'. While much interesting work has been done in the 30 years since this statement was made, behavioural biomarkers have still not reached the stage where they are accepted as part of formal testing procedures.

There are two fundamental difficulties facing the use of behavioural tests in wildlife toxicology, first that the best studied and most easily performed and quantified are those that have the least environmental relevance, and second that the most relevant behaviours are the most strongly conserved against change.

Operant behaviour, such as conditioning to respond to a coloured key to obtain food, is too remote from real life to be capable of being related to survival. It can merely be presumed that a decrease in learning ability is an unfavourable response. Avoidance behaviour is more directly related to survival, although the relationship has not been quantified. The ability to capture food is clearly important to predatory species, but is difficult to measure under field conditions.

Field observations are usually difficult to quantify. It was suggested that behavioural changes were a possible cause in the decline of the peregrine falcon. However, observations by time-lapse at the eyries of highly contaminated peregrines revealed little in the way of abnormal behaviour. This study was based on seven peregrine eyries in Alaska, using battery-powered time-lapse motion pictures cameras, taking pictures about every 3 minutes and the film cartridges needed to be replaced every 6–7 days. Even so it was a full-time job to replace the film cartridges as the eyries were widely separated and the terrain was difficult. In two of the nests the eggs broke, but no evidence of abnormal behaviour was observed. The other five nests were successful. In all some 70 000 pictures covering 4200 h were obtained. One of the drawbacks of this type of experiment is the time taken to analyze the data obtained. In another study (Nelson, 1976) observations from a blind, totalling over 300 h, were made on 12 peregrine eyries. Four clutches lost single eggs, probably by breakage, but no abnormal behaviour was observed. Although 300 h is a lot of time to spend sitting in a blind, it is only 25 h per clutch out of 400 h of daylight during the incubation period.

These two studies illustrate the difficulties of making field observations on behavioural changes. An additional problem is the fact that even if behavioural changes are documented it is difficult to relate them to a specific chemical or chemicals. The best-documented studies that can be extended to real-life situations are those involving the organophosphorus pesticides. Here there is a well-defined biochemical biomarker (inhibition of AChE) that can used in conjunction with behavioural changes, and an extension to field situation can be made (Chapter 15).

A considerable number of studies of the behavioural effects caused by OP pesticides have been made. With these pesticides the reference line is the inhibition of AChE. The relationship of a variety of behavioural changes to degree of AChE inhibition has been determined for several avian species. Some of these are listed in Table 10.4. Although a variety of behaviours are altered, these only occur at substantial inhibition of AChE suggesting the measurement of this enzyme is an easier and more sensitive assay. However, as discussed in Chapter 15, behavioural changes affect the selection of organisms collected for AChE assay. An interesting study under operational conditions was that of interaction of predation pressure and the effect of the insecticide abate on populations of fiddler crabs. In these studies the population densities of crabs in open marshes treated and untreated with abate and in plots that were caged over and those that were open were measured. It was found that the population of crabs in the caged areas was similar in both treated and untreated areas, but that following the second application of abate there was a significant decrease of the population of crabs in the uncaged treated areas. Experimental studies suggest that the escape response of the crabs was impaired.

Table 10.4 Relationship of behavioural effects to cholinesterase inhibition[a]

Species	Pesticide	Degree of inhibition	Effects seen
Laughing gull (*Larus atricilla*)	Parathion	50%	Time spent incubating eggs was decreased
		46%	No change in flushing distance or return time to nest
American kestrel (*Falco sparverius*)	Acephate	25%	Feeding habits and attack behaviour not affected
Bobwhite quail (*Colinus virginianus*)	Methyl parathion	57%	Increased predation by cats
Starling (*Sturnus vulgaris*)	Dicrotophos	50%	Increased time perching, decreased flying, foraging, singing and displaying; females spend more time away from nest

[a] From Peakall (1985).

Studies of avoidance responses of fish to toxicants dates back over 80 years. From the first simple studies on acetic acid the field has grown enormously and the equipment used has become highly complex. We now have sophisticated fish avoidance chambers with video monitor and computer-interfaced recording systems.

Recent studies include many on the effects of heavy metals. If one compares the lowest-observed-effect concentration (LOEC) obtained from behavioural studies (avoidance, attractance, fish ventilation and cough rates) with chronic toxicity studies one finds that some of the behaviour tests are more sensitive than life cycle or early life stage tests. Other tests involving predator avoidance, feeding behaviour, learning, social interactions and a variety of loco-motor behaviours have been insufficiently studied to enable a judgment of their sensitivity or utility. At the moment, behavioural tests have not replaced conventional toxicity tests. However, behavioural tests may provide ecological realism, for example the effects of pollutants on predator–prey relationships. Such tests must be capable of field validation.

Behavioural tests that are most advanced are those involving fish. The fish avoidance test is well established in the laboratory as a means of showing effects well below the lethal range and highly automated procedures are available. Nevertheless, a note of caution should be injected. It has been found that pre-exposure to effluent reduced the avoidance behaviour, and pre-exposed fish were observed more often in contaminated than in clean water (Hartwell *et al.*, 1987). This desensitization caused by pre-exposure make it likely that laboratory experiments will overestimate the responsiveness of fish to metal pollution in the wild.

These difficulties do not imply that behavioural effects caused by pollution are unimportant. Studies such as those on fiddler crabs have shown that operational levels of pesticide use can cause population effects through behavioural changes. Sub-standard prey are more readily captured by predators, but field studies of the impact of chemicals on behaviour are difficult.

The overall conclusion is that behavioural parameters are not especially sensitive to exposure to pollutants and that biochemical and physiological changes are usually at least as sensitive. Further, the variability of biochemical data are generally less and the dose response clearer than those obtained from behavioural studies. In general, physiological and biochemical changes are more readily measured and quantified.

10.4.6 Biomarkers in Plants

Plants have widely been used as biomonitors to localize emission sources or to analyze the impact of pollutants, especially gaseous air pollutants, on plant performance, one of the earliest ones being by Angus Smith who coined the term 'acid rain' after examining the damage done to plants around Manchester in the middle of the last century.

However, biomarkers should go beyond the visible parameters of sentinel species. They should establish such processes and products of plants, which enable a recognition of environmental stress earlier than visible damage. Biomarkers must therefore be able to predict the environmental outcome and consequential damage. Ideally, biomarkers should be selected from the events of biochemical or physiological pathways, but at the present time we have not yet reached this stage with plant biomarkers.

Specific biomarkers have been identified in sensitive plants. In a few cases it is known that excess of a specific chemical will give rise to the production of a metabolite which is different between tolerant and sensitive plants.

In the presence of an excess of selenium, Se-sensitive plants fail to differentiate between S and Se. They incorporate Se in sulphur amino acids leading to the synthesis of enzymes of lower activity which can lead to plant death. In contrast, Se-tolerant plants biosynthesize and accumulate non-protein seleno amino acids such as selenocystathioneine and Se-methylselenocysteine which do not cause metabolic problems for the plant. Thus, the occurrence of seleno proteins in plants are excellent biomarkers for Se stress although their use in the field has not been widely reported.

Another example is that after an exposure to an excess of fluor (a mineral containing fluorine), plants synthesize fluoroacetyl-CoA, and then convert it, via the tricarboxylic acid cycle, to fluorocitrate. The latter compound blocks the metabolic pathway by inhibiting the enzyme aconitase. As a result of this process, fluorocitrate accumulates and is a very reliable biomarker for fluor poisoning.

There are some plant biomarkers for free heavy metals. Phytochelatins are synthesized during exposure to a number of heavy metals and anions as SeO_4^{2-}, SeO_3^{2-}, and AsO_4^{3-}. Dose and time-dependent relationships have been established under laboratory conditions for cadmium, copper and zinc. For monitoring purposes, research is needed into the phytochelatin production of plants in the field.

General plant biomarkers, which respond to a variety of environmental stresses, may be useful to indicate that something in the environment is a hazard to plant life. For example, the activity of the enzyme peroxidase has been used to establish the exposure of plants to air pollution, especially SO_2.

Changes of enzyme systems during the development stage of the plant, as well as the effects of seasonal and climatic processes, are not yet well enough known for the activity of enzymes to be reliable as monitoring devices. At present, plant biomarkers are not as well advanced as animal biomarkers. However, this situation is likely to change. The fact that plants are stationary aids greatly in measuring exposure to pollutants; monitoring surveys for measuring levels of heavy metals in lichens and organochlorines in pine needles are in place and would be valuable if these measurements could be linked with biological changes within the plant.

10.5 Role of Biomarkers in Environmental Assessment

The most compelling reason for using biomarkers in environmental assessment is that they can give information on the effects of pollutants. Thus, monitoring the use of biomarkers is complementary to the more usual monitoring involving the determination of residue levels. The first point in any assessment process is to decide what is being assessed. This may sound self-evident, but it is surprising how seldom, precise objectives are defined. This is the case with many monitoring programmes to determine the levels of environmental chemicals. Take, for example, the International Mussel Watch Programme which measures heavy metals in mussels in many parts of the world. The justification of such surveys is that we should know what pollutants are, where and at what concentration. However, what is going to be done with the information? Only if it is known what concentrations are hazardous and if effective remedial action can be taken will this information be of practical use. Similar considerations apply to the use of biomarkers. Again, action levels have to be decided if the monitoring programme is to be effective. An advantage of the biomarker approach is that it may show that the physiology of the organisms is within normal limits indicating that no action is necessary. By contrast, zero levels are rarely found in analytical determinations. Ideally both approaches (i.e. residues and biomarkers) should be used together in an integrated manner (Chapter 11).

Hazard is a function of exposure and toxicity (Chapter 6). In the simplest terms, if there is no exposure there is no hazard and if there is no toxicity there is no hazard. Legislation, such as EEC 6th and 7th Amendments refer, in general terms, to protecting man and the environment. This legislation has specific requirements for data that must be provided at various stages of production. There is a basic set – called the minimum premarket data – that must be provided for all new chemicals before any production occurs. Increasing amounts of data must be produced as the amount of the chemical produced increases. The use to which these data are put in hazard assessment is not spelled out in the EEC legislation. Rather this is left to individual governments. A comparison of the various approaches that have been used in prioritization and standardization of hazard assessment are considered by Hedgecott (1994).

Before legislation on limiting hazard can be enforced, two fundamental questions must be answered: (1) how much damage are we prepared to tolerate? and (2) how much proof is enough? At one extreme there will not be much concern if a few aquatic invertebrates die within a few metres of the end of an outlet pipe. At the other end of the scale, an event such as the destruction of most of the biota of the Rhine (Deininger, 1987) resulted in a worldwide reaction. Our concerns are also species-dependent and tend to increase as we move from algae to mammals. Even within a class of animals there are widespread differences: it is easy to arouse concern over pandas and whales, difficult in the case of rats. Further, some types of damage are considered more serious than others, alteration of the genetic material, which may be passed onto future generations, being considered one of the most serious effects.

The question 'how much damage are we prepared to tolerate?' is a question for society to answer. Once it is answered it is possible to design protocols to meet the standards required. Unanswered, it leaves the scientists with the problem of trying to set up regulations in an impossible situation.

The second question 'how much proof is enough?' is largely a scientific one, although it requires the first question to be answered before it can be tackled. There is considerable inertia in decision-making, but decisions must be made. 'No decision' is a decision.

Why use biomarkers in hazard assessment? One important reason lies in the limitations of classical hazard assessment. The basic approach of classical hazard assessment is to measure the amount of the chemical present and then relate that, via animal experimental data, to the adverse effects caused by this amount of chemical. The limitation of this approach is that only for a very few compounds has it been possible to define the levels of a chemical that are critical to an organism. Under real life situations a wide variety of organisms are exposed to complex and changing levels of mixtures of pollutants. Further chemical monitoring only works if the material is persistent. Chemicals such as the PAHs and many pesticides have very short biological half lives in most species but may, nevertheless, have long-term effects. Biological and chemical monitoring systems should be complementary to each other. It is important to know both what is there and what it does.

The first question that biomarkers can be used to answer is 'are environmental pollutants present at a sufficiently high concentration to cause an effect?'. If the answer is positive, further investigation to assess the nature and degree of damage and the causal agent or agents is justified. If negative, it means that additional resources do not have to be invested. In this way biomarkers can act as an important 'early-warning' system.

The role of biomarkers in environmental assessment is envisaged as determining whether or not, in a specific environment, organisms are physiologically normal. The approach has similarities to the use of clinical biochemistry in human medicine. A suite of tests can be carried out to see if the individual is healthy. It is necessary to select both the tests and the species to be tested. The selection of indicator species is going to be, at least to some extent, site specific. However, there is merit in having as much commonality of species as is feasible between studies. It is important to see that the main trophic levels are covered and not to rely completely on organisms at the top of the food chain. Although these have been the species most affected by some pollutants such as the persistent organochlorines, there is no reason to believe that this will be a universal truth.

In the selection of tests the specificity of the test to pollutants and the degree to which the change can be related to harm need to be considered. Both specific and non-specific biomarkers are valuable in environmental assessment. In an ideal world we would have biomarkers to indicate the exposure to, and to assess the hazard of, all major classes of pollutants and non-specific biomarkers that assess accurately and completely the health of the organism and its ecosystem.

From the above it is clear that there are two main aspects to a definition of harm, one scientific and the other social. Scientifically it is important to demonstrate unequivocally that changes have occurred as a result of pollution. But whether or not that change is sufficiently serious to make it essential to bear the cost of the remedial action is something for society to decide.

Further Reading

FOSSI, M. C. and LEONZIO, C. (eds) (1993) *Nondestructive Biomarkers in Verte-brates* Proceedings of a workshop devoted to the use of blood and other tissues that can be collected non-destructively in the measurements of biomarkers.

HUGGETT, R. J. *et al.* (1992) *Biomarkers. Biochemical, Physiological and Histological Markers of Anthropogenic Stress* In addition to those techniques listed in the title there is extensive coverage of DNA alterations and immunological biomarkers.

MCCARTHY, J. F. and SHUGART, L. R. (1990) A compilation of a large number of papers presented by both American and European scientists at an American Chemical Society meeting.

PEAKALL, D. B. (1992) *Animal Biomarkers as Pollution Indicators* A single-author work looking at the use of biomarkers in higher animals in environmental assessment.

PEAKALL, D. B. and SHUGART, L. R. (eds) (1993) *Biomarkers. Research and Application in the Assessment of Environmental Health* NATO workshop that looks at the strategy of using biomarkers in environmental assessment.

11

In Situ Biological Monitoring

11.1 Introduction

It is difficult to predict the effects of pollutants on organisms to an acceptable degree of accuracy by simply measuring concentrations of a chemical in the abiotic environment (Figure 11.1). Factors which affect bioavailability of chemicals to organisms include temperature fluctuations, interactions with other pollutants, soil and sediment type, rainfall, pH and salinity. Even using

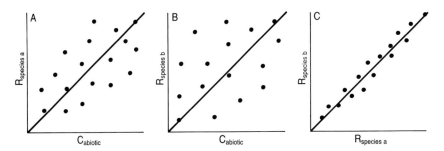

Figure 11.1 Schematic graphs to illustrate a principal of *in situ* biological monitoring of pollution. In this hypothetical example, the responses, R (as measured by concentrations or effects of the pollutant on the y axis), of species 'a' (A) and 'b' (B) in sites with different levels of contamination, are not closely related to concentrations of the pollutant (x axis) in abiotic samples (soil, air, water sediment) from the same sites. Because the relationship between species in the same sites is much closer (C), the responses of species 'a' to the pollutant can be used to predict the responses of species 'b' more accurately than similar predictions from abiotic samples (see Figures 11.5 and 11.6 for data that support this hypothesis). Reproduced from Hopkin (1993a) with permission of Blackwell Scientific.

biotic monitoring of chemical residue levels (see section 11.3) there are considerable difficulties in knowing the effects of this level of chemical on the organism. This process is made more difficult by the presence of mixtures (see Chapter 10) and the considerable inter-species differences in response. *In situ* biological monitoring attempts to get around these problems by analyzing various parameters of natural populations which reflect the situation in the field rather than the standardized conditions of laboratory experiments.

There are four main approaches to *in situ* biological monitoring of pollution (Hopkin 1993a). Each will be examined in detail in this chapter and will be illustrated with examples of recent studies on a range of animals and plants from terrestrial and aquatic ecosystems. The four involve:

1. monitoring the effects of pollution on the presence or absence of species from a site, or changes in species composition, otherwise known as 'community effects' (see also Chapter 12);
2. measuring concentrations of pollutants in indicator or 'sentinel' species (see also Chapters 4 and 5);
3. assessing the effects of pollutants on organisms and relating them to concentrations in those organisms and other biotic and abiotic indictors (see also Chapter 9);
4. detecting genetically different strains of species which have evolved resistance in response to a pollutant (see also Chapter 13).

11.2 Community Effects (type 1 Biomonitoring)

The most frequent response of a community to pollution is that some species increase in abundance, others (usually the majority) decrease in abundance and populations of others remain stable. The patterns of the species abundances reflect effects integrated over time and are used widely to monitor effects of pollutants on communities.

11.2.1 *Terrestrial Ecosystems*

In order to be able to recognise an unusual assemblage of species, it is necessary to monitor changes over time, or have sufficient background knowledge of the 'normal' ecology of similar but unpolluted sites. In the former approach, sites must be monitored for several years to distinguish effects of pollutants from natural fluctuations. Perkins and Millar (1987) showed that emissions from an aluminium works on Anglesey, North Wales, were responsible for the almost complete elimination of lichens within 1 km of the factory soon after its opening in 1970. Some recovery has taken place since 1978 when new emission controls were introduced, but the lichen flora is still very impoverished in

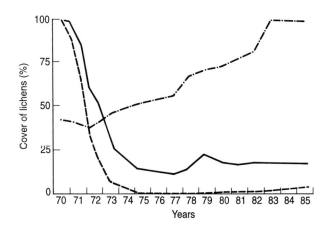

Figure 11.2   Cover (as percentage of type maximum) of fruticose (○), foliose (●) and crustose (△) coricolous lichens in eight permanent quadrats, set up on broadleaved trees within 1 km of the aluminium works in Anglesey, North Wales, during 1970–1985. Reproduced from Perkins and Millar (1987) with permission of Elsevier Applied Science.

comparison to pre-1970 populations (Figure 11.2). In the latter 'background knowledge' approach, one or more sites are examined in the contaminated area and are compared with at least one 'reference' site (e.g. Figure 11.3). However, if the differences between polluted and reference sites are subtle, the investigator is left with the difficult problem of deciding when a change indicates a toxic effect or natural between-site variations.

11.2.2   *Freshwater Ecosystems*

In Britain, three main approaches have been adopted to assess the effects of pollution on communities of freshwater organisms (British Ecological Society, 1990, on which this section is based). There are:

1. the *biotic* approach, based on the differential sensitivities of species to pollutants;
2. the *diversity* approach, based on changes in community diversity;
3. *River Invertebrate Prediction and Classification* (RIVPACS), which combines an assessment in terms of both the types of species present and the relative abundances of families.

The most frequently used biotic indices have been the Trent Biotic Index (TBI), Chandler Biotic Score (CBS) and Biological Monitoring Working Party (BMWP). All three are based largely on relative tolerances of macro-invertebrates to organic pollution. TBI and CBS require identification of

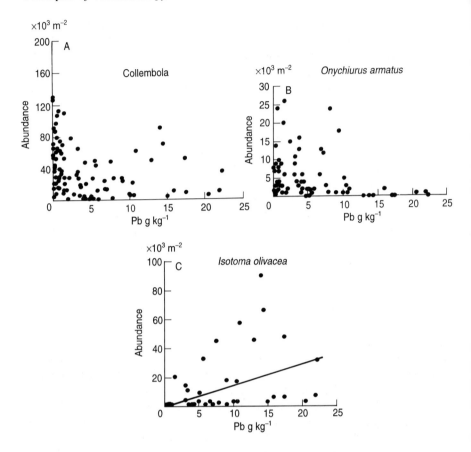

Figure 11.3 Abundance of (A) total Collembola, (B) *Onychiurus armatus* (Collembola) and (C) *Isotoma olivacea* (Collembola) in the 0–3 cm layer in lead-contaminated soils in the vicinity of a natural metalliferous outcrop in a Norwegian spruce forest. The concentrations of lead represent metal extracted from soil over 18 h in 0.1 M buffered acetic acid. Note that *O. armatus* is sensitive to lead pollution whereas *I. olivacea* reached higher population densities in contaminated soils. Reproduced from Hågvar and Abrahamsen (1990) by permission of the Entomological Society of America.

species whereas BMWP requires only family level identification. Only the CBS takes abundance into account. However, these scores are generally assumed rather than experimentally derived. Furthermore, sensitivity rankings are assumed to apply across a range of toxicants even though laboratory experiments have shown that this is not necessarily the case.

The most frequently used diversity indices have concentrated on species richness and the distribution of individuals among species. The *Shannon–Weiner diversity index* is the most commonly used. However, there are three main problems with diversity indices. First, many factors influence community structure and as diversity indices do not take account of the species present, their usefulness as a measure of water quality can be questioned. Second, there is an unresolved debate as to which index to use to measure diversity and as to which taxonomic group and level should be considered. Third, it is not clear how diversity responds to pollution. For example, diversity of plankton reduces continuously with organic enrichment, but for benthic invertebrates the response in 'bell-shaped' with the greatest diversity at intermediate pollution levels (British Ecological Society, 1990).

The RIVPACS approach is used to predict the fauna of a site using environmental variables (Wright *et al.*, 1993). Hypothetical target communities are provided against which the combined effects of physical and chemical stresses can be assessed. Comparison of observed values with these predictions provides environmental quality indices. Although RIVPACS is the most widely used of the techniques described, it is still a 'broad brush' approach more useful for highlighting rivers and streams in need of more detailed study than giving a final verdict on the level of environmental damage.

11.2.3 *Marine Ecosystems*

One of the simplest methods of detecting a pollution-induced change in communities of marine benthic organisms is to analyze the log normal distribution of individuals per species in sediment samples (Gray, 1981). In many samples of benthic communities, the most abundant class is not that represented by one individual per species but often lies between classes with three and those with six individuals per species. Thus the curve relating numbers of individuals per species (x axis) to number of species (y axis) is often strongly skewed. This curve can be 'brought back' to a normal shape by plotting the number of individuals per species on a geometric scale (usually X2). Plotting the geometric classes on the *x* axis (class I = 1, class II = 2–3, class III = 4–7, class IV = 8–15, and so on) against cumulative percentage of species on the y axis invariably gives a straight line. In polluted sites, there is often a break in the line indicating a departure from an equilibrium community. If this persists over several sampling occasions, it is indicative of pollution-induced disturbance.

The responses of marine-fouling communities to pollution stress, in terms of changes in species composition, can be monitored *in situ* by reciprocal transplants. Climax communities are allowed to develop on submerged surfaces in a clean and a polluted site, and are then moved between the sites. An experiment in Australia at Woolongong Harbour (uncontaminated) and Port Kembla Harbour (polluted by discharges from heavy industry), demonstrated rapid changes in the community structure in response to pollution (Moran and Grant, 1991). Indeed within two months, those communities on submerged plates that had been transferred from Woolongong to Port Kemblar were similar in structure to those that had developed entirely in Port Kemblar.

Most changes occurred in short time periods when sensitive species were killed by periodic discharges (an effect difficult to predict by measuring levels of pollutants in the water). Space previously occupied by these species was quickly colonized by opportunists more tolerant to the pollutants, thus leading to changes in community structure.

11.3   Bioconcentration of Pollutants (Type 2 Biomonitoring)

Destructive measurement of levels of pollutants in organisms provides an indication of how much is present at a particular moment in time and may enable effects on predators to be assessed (see Chapters 12 and 15). Take, for example, a species of wading bird that feeds primarily on estuarine bivalve molluscs. 'Critical' (safe?) concentrations for the birds could be set based on levels in bivalves rather than sediment or water, or indeed their own tissues. The bivalves provide a critical pathway from the abiotic environment to the waders, the importance of which can be monitored biologically by analyzing the molluscs.

11.3.1   *Terrestrial Ecosystems*

Contamination of plants has been monitored either by collecting samples directly from the field, or exposing material (usually bags of *Sphagnum* moss) for a specified period and returning it to the laboratory for analysis. The recent decline in concentrations of lead in air in the UK following the reduction of permitted levels of lead in petrol has been mirrored by a decline in lead concentrations in plants (Jones *et al.*, 1991).

Biological monitoring of radioactive fall-out in Italy derived from the Chernobyl disaster in 1986 has shown that levels of ^{137}Cs in mushrooms have been increasing since the accident (Borio *et al.*, 1991). Basidiomycete hyphae in the soil accumulated the radioisotopes but it was not until they produced sporophores that the caesium became available to above-ground fungivores. No correlation was found between the level of ^{137}Cs in the mushrooms and the soil in this study, emphasizing the importance of biological monitoring of radioactive pollution. Traditional risk assessment does not make allowances for effects such as this.

In Norway, where similar results have been obtained, the level of ^{137}Cs in the milk of goats increased five-fold in 1988 following abundant growth of mushrooms in grazing land (Figure 11.4). The sporophores contained levels of radioactivity up to 100 times those in green vegetation. Thus, fungi provide an important critical pathway for the concentration and transport of radioactive isotopes along food chains.

As far as invertebrates are concerned, it is clear that some groups, such as woodlice, snails and earthworms, accumulate significant amounts of metals

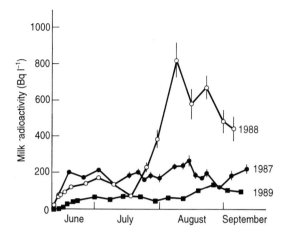

Figure 11.4 Radiocaesium (^{134}Cs and ^{137}Cs) activity from 1987 to 1989 in goat milk during the grazing seasons from 15 June to 15 September in the Jotunheimen mountain range (mean + SE, three to eight animals in 1987 and eight in 1988 and 1989). Reproduced from Hove *et al.* (1990) with permission of the Health Physics Society and Pergamon Press.

from their diet (Hopkin, 1989) whereas most insects are able to regulate concentrations to relatively low levels (Hopkin, 1995). Thus, species in the former three groups provide a significant route for the transfer of metal pollutants to their predators (Figures 11.5, 11.6).

Monitoring concentrations of pollutants in vertebrates may pose some difficulties. They are usually difficult to catch, population densities are lower than in invertebrates, and a licence may be required (or may even be unobtainable for some species). Ways around this are to take blood samples, or use a non-living product of the animal which indicates previous exposure to a pollutant. Analysis of eggs has been widely used for organochlorines and this is less destructive than the collection of adults. Feathers have been proposed as possible indicators. In Finland, feathers from nestlings of a range of birds provide a good indicator of mercury exposure of the adults (Solonen and Lodenius, 1990). Regurgitated pellets from owls can be used to monitor exposure to rodenticides.

11.3.2 *Freshwater Ecosystems*

Predicting bio-accumulation in aquatic systems is more difficult than in terrestrial ones due to the greater mobility of water and sediments in comparison to soils, and the difficulty of knowing whether the main route of exposure is via water or food. Organisms can be collected directly from the field for analysis,

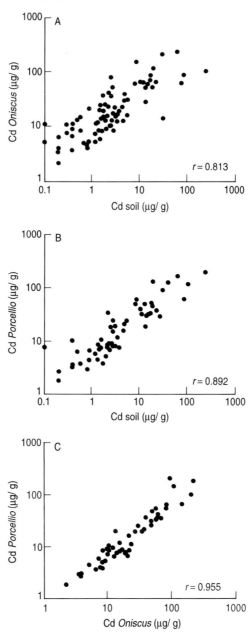

Figure 11.5 Relationships between concentrations of cadmium (dry weight) in the terrestrial isopods (woodlice) *Oniscus asellus* (A) and *Porcellio scaber* (B) and soil, and between the two species (C) collected from sites in Avon and Somerset, SW England in 1988 and 1989. The region includes a primary zinc, cadmium and lead smelting works, and disused zinc mining areas. Each point represents the mean of 12 isopods and six samples of soil from each site. Note that the concentrations of cadmium in *P. scaber* in this region can be predicted more accurately from the concentrations in *O. asellus* (C) than from levels in soil (B). Reproduced from Hopkin (1993a) with permission of Blackwell Scientific.

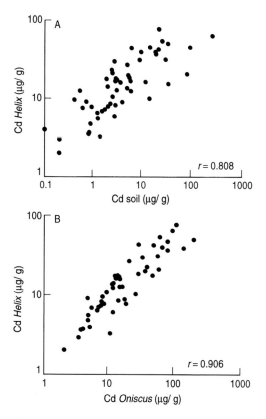

Figure 11.6 Relationships between concentrations of cadmium (dry weight) in the snail *Helix aspersa* and soil (A) and *Oniscus asellus* (B) collected from the same region as Figure 11.5. Each point represents the mean of seven snails, 12 woodlice or six samples of soil from each site. Note that the concentrations of cadmium in *H. aspersa* can be predicted more accurately from the concentrations on *O. asellus* (B) than from levels in soil (A). Reproduced from Hopkin (1993a) with permission of Blackwell Scientific.

or caged in polluted and unpolluted sites to assess bio-availability. The fresh-water amphipod *Gammarus pulex* has been used extensively for such work.

Monitoring concentrations of pollutants in fish is carried out all over the world. For example, a monitoring programme for mercury in Brazil showed that levels in edible parts of fish from gold mining areas (where mercury is used extensively in gold extraction and refining) were five times greater than the 'safe' level for human consumption (Pfeiffer *et al.*, 1989). The long-term studies of Schmitt and Brumbaugh (1990) detected a decrease in the concentrations of lead in fish in the USA between 1976 and 1984. This coincided with regulatory measures that reduced the influx of lead to the aquatic environment. A similar downward trend has been found also in the levels of PCBs in the eggs of herring gulls from the Great Lakes (Figure 11.7).

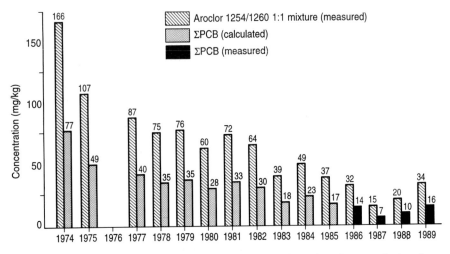

Figure 11.7   Concentrations of PCB in eggs of herring gulls from the adjacent colonies of Muggs Island (1974–1986) and Leslie Street Spit (1987–1989), Lake Ontario. Calculated concentrations for $\sum$PCB obtained from Aroclor 1254/1260 1 : 1 mixture were obtained using the conversion factor for Lake Ontario. Reproduced from Turle *et al.* (1991) with permission of Pergamon Press.

11.3.3   *Marine Ecosystems*

The effects of pollutants on marine ecosystems are difficult to study because of the vastness of the system, problems in obtaining specimens and the difficulty of relating any effects seen to specific chemicals. Marine mammals at the top of the food chain are particularly difficult as they are usually protected, so that sampling is difficult and laboratory studies to establish effects are few. Detailed studies on the effects of PCBs on seals have been made in The Netherlands (Brouwer *et al.*, 1989), but this is an exception and usually extrapolation has to be made from data from entirely different species.

By far the greatest research effort has been directed towards bivalve molluscs. The reasons for this are four-fold. First, many species are a source of food for predatory vertebrates, particularly birds. Second, they are widespread and common, are easily collected in large numbers and are sedentary. Third, because they are filter feeders, bivalves pass large volumes of water through their bodies, accumulate pollutants continuously and act as integrators of exposure over long periods. Fourth, there is good background knowledge of their basic biology to allow results to be put into an ecotoxicological context.

Mytilus edulis has been analyzed most frequently since it is common and has a global distribution. Research on *Mytilus* has been so extensive that global 'mussel watch' schemes have been established. These have shown trends in pollutant levels which, in many cases, have declined. For example, Fischer

(1989) was able to demonstrate that concentrations of cadmium in *Mytilus edulis* in Kieler Bucht in the western Baltic in 1984 had declined to about 30% of their level in 1975. Many of the schemes are ongoing and have been running for several years. This will ensure that long-term trends in the bioavailability of inorganic and organic pollutants to bivalves and their predators can be separated from natural fluctuations.

Vertebrates are at the top of marine food chains and are vulnerable to poisoning from pollutants contained in their diets. This is particularly true of some organic pollutants which may accumulate in fatty tissues of marine mammals and birds. Residence times may be very long in organic pollutants which are highly lipophilic ($K_{Ow} > 10^5$), or which are only slowly metabolized to water-soluble products that can be excreted (see Chapter 5). Thus, even after the complete removal of the source of pollution, contaminant levels in the tissues of vertebrates may take several years to decline to background levels. Levels of PCBs and DDT in open ocean dolphins did not decline between 1978 and 1986 despite a reduction in organochlorine contamination of the marine environment during this period (Loganathan *et al.*, 1990).

11.4   Effects of Pollutants (type 3 Biomonitoring)

The primary aim of ecotoxicologists should be to describe and predict *effects* of pollutants on organisms and ecosystems. The basis of such studies is that biochemical, cellular, physiological and morphological parameters can be used as screening tools or 'biomarkers' in environmental monitoring (see Chapter 9).

11.4.1   Terrestrial Ecosystems

One of the simplest *in situ* indicators of air pollution is the Bel W3 variety of tobacco. The plants develop a mottling of the leaves in response to low levels of ozone pollution. The test is easy to perform and has been used in several surveys involving schoolchildren (Heggestad, 1991). Forest dieback is one of the clearest examples of the effects of pollution on ecosystems. Growth rates of trees can also be useful and can be retrospective if determined from the widths of annual growth rings.

In some situations, effects can be related directly to a local source of pollution. Walton (1986) obtained direct evidence for the effects of fluoride emissions from an aluminium plant on small mammals. Moles and shrews collected less than 1 km from the plant had extremely high levels of fluoride in their bones and teeth. Several manifested the symptoms of fluoride poisoning including chipped and broken teeth and brittle bones.

An *in situ* bioassay using earthworms for assessing the toxicity of pesticide-contaminated soils was developed by Callahan *et al.* (1991). At each site, five

Lumbricus terrestris were placed in enclosures distributed in transects through-out areas of high and low contamination at a 'superfund' site in Massachu-setts, formerly used for mixing pesticides. Mortality, morbidity (coiling, stiffening, swelling, lesions etc.) and whole body concentrations of a wide range of organic pollutants in worms were related to levels in the soils. This *in situ* method does not require removal of highly contaminated soils from the site. It provides an accurate dose–response relationship which can be used to predict the soil levels below which worms will return to the site, and the 'safe' levels in soils at which worms will not accumulate sufficient concentrations of pol-lutants to be harmful to their predators.

11.4.2 Freshwater Ecosystems

At the organism level, most recent research has been directed towards developing *in situ* bioassays for detecting sublethal effects. The most widely-used parameter is 'Scope for Growth' (SFG). SFG measures the difference between energy input to an organism from its food and the output from respir-atory metabolism and, at least in principal, can be related to population and community processes. Animals which are 'stressed' (expending energy on detoxifying and excreting pollutants) have less energy available for somatic growth and reproduction. This is manifested as lower reproductive and growth rates compared to unstressed controls. Most measurements on SFG in fresh-water have been on amphipod and isopod crustaceans. The SFG test using *Gammarus pulex* for measuring the effects of zinc and low pH is at least an order of magnitude more sensitive than acute 24-h LC_{50} tests (Naylor *et al.*, 1989).

Many rivers and streams are affected by acute episodic pollution rather than long-term chronic contamination. 'Bursts' of pollutant run-off can occur after thunderstorms, rapid snow melt (see Figure 4.8), accidents at factories or deliberate release. These effects of intermittent increases in pollutant concen-trations can be detected by continuous monitoring of caged organisms. Seager and Maltby (1989) described such a system which used rainbow trout (*Salmo gairdneri*). Fish were caged *in situ* in Pendle Water, a polluted urban river in Lancashire, England. The trout responded to sewer outflow discharges by increasing their breathing rate. This was monitored by measuring the small oscillating voltage produced by the muscles involved in gill ventilation.

11.4.3 Marine Ecosystems

A classic example of the effects of pollutants on marine organisms is that of tributyl tin (TBT) leached from antifouling paints. These are applied to the hulls of boats to inhibit settlement of marine invertebrates, particularly barnacles and bivalve molluscs. A summary of the effects that TBT can have is given below. The topic is also covered in Chapter 9 and 15.

A combination of traditional laboratory toxicity studies, transplant experiments and field observations has shown that TBT is probably the most toxic synthetic substance ever to have been released deliberately into the marine environment (Bryan *et al.*, 1988). TBT affects many marine organisms and has had a severe economic impact on oyster fisheries and farms. In British waters, the chemical has had a dramatic effect on the dog whelk *Nucella lapillus* which is hypersensitive to TBT. Dog whelks are now absent from many sites in the UK where they were common before the introduction of TBT paints (Bryan *et al.*, 1986).

TBT causes female dog whelks to grow a vas deferens and a penis. These block the opening of the female genital duct so that eggs cannot be released (Gibbs and Bryan, 1986). TBT seems to affect the hormone system that determines the sex of prosobranch molluscs (Gibbs *et al.*, 1988). Similar findings have been made on other *Nucella* species elsewhere in the world.

The mean size of the female penis relative to males in a population of dog whelks is called the level of 'imposex'. The level is calculated by dividing the cube of the mean length of the female penis by the cube of the mean length of the male penis (both in mm) and multiplying by 100. Thus, if the level of imposex is 100% then the mean size of the female penis is the same as that of the males. The imposex index provides an indicator of the exposure of dog whelks to TBT at a site (Figure 11.8).

Work on the beluga whale (*Delphinapterus leucas*) in the St Lawrence estuary in Canada illustrates the difficulties of field studies. This population

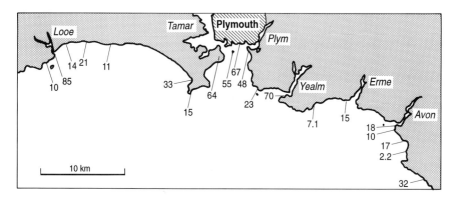

Figure 11.8 Levels of 'imposex' (percentage size of female penis relative to males) in dog whelks (*Nucella lapillus*) collected from SW England in 1984–1985. Imposex develops in females in response to TBT leached from anti-fouling paints and is most prevalent in areas of high boating activity (e.g. the Looe and Yealm estuaries). Reproduced from Bryan *et al.* (1986) by permission of the Marine Biological Association of the United Kingdom and Cambridge University Press.

has failed to increase despite the cessation of hunting and toxic chemicals have been blamed for the problems of this isolated population. Here, collection of specimens for scientific study cannot be made and thus material can only be obtained from animals found dead, or by non-destructive sampling. Determination of the residue levels revealed high levels of PCBs but since the animals had been dead for some time, measurements of the biomarkers studied by the Dutch workers were not possible. At the present time there is no firm evidence that PCBs are causing effects in whales.

It should be pointed out that although both pinnipeds and cetaceans are termed 'marine mammals', they are not closely related. Studies on DNA adducts (sections 7.3.1 and 9.4.3.1) from stranded whales in the St Lawrence have revealed high levels of benzo(a)pyrene adducts which were not detected in the brains of belugas killed by native hunters in the Arctic (Martineau *et al.*, 1988). The levels of adducts have been correlated with high incidence of tumours in the St Lawrence beluga whales.

The possibility that the widespread mortality of seals in various parts of the world caused by viruses is linked to effects of chemicals (especially PCBs) on the immune system has been put forward, but the Scottish verdict of 'not proven' seems to be as far as one can go at the moment. Detailed immunological studies on beluga whales have been started based on non-destructive sampling of skin and it will be interesting to learn if the high levels of PCBs in the St Lawrence population is affecting the immune system.

11.5   Genetically-based Resistance to Pollution (type 4 Biomonitoring)

Strains of plants which possess genetically-based resistance to high concentrations of metals in soils have been recognised for many years (Baker and Walker, 1989). Resistance is also well-documented in insects. Such resistance is inheritable and should be distinguished from phenotypic tolerance which all members of a species may possess ('pre-adaptation'). The latter may consist of avoidance strategies, high excretion ability, or possession of enzymes that break down organic pollutants. Phenotypic tolerance can be induced (e.g. increased synthesis of metal-binding proteins). However, genetically distinct, pollution-resistant strains will evolve only if the selection pressure persists for several generations.

Terrestrial invertebrates which have been shown by breeding experiments to be genetically resistant to high concentrations of metals include races of Collembola, terrestrial isopods (woodlice) and *Drosophila* (Posthuma and van Straalen, 1993). Freshwater oligochaetes and marine polychaetes have also evolved resistance to metals. In some of these cases, the basis of the increased tolerance is an increase in the copy number, or transcription rate of the gene coding for detoxifying proteins. In the case of organophosphate insecticides, up to a 256-fold amplification of the genes coding for non-specific esterases that break down the insecticide have been found in mosquitoes (see also

Figure 8.12). For metals such as copper and cadmium, the gene that codes for the metal-binding protein metallothionein may be duplicated up to four times. The metals are bound more rapidly by resistant animals following ingestion. Such amplification has been found in wild *Drosophila* and has probably evolved in response to the spraying of fruit trees with copper-containing fungicides.

11.6 Conclusions

In situ biomonitoring organisms should satisfy the '5Rs' if they are to be used successfully (Hopkin, 1993a). These are:

1. *Relevant* – to be ecologically meaningful, ecotoxicological tests should use species that play an important role in the functioning of the ecosystem.
2. *Reliable* – species should preferably be widely distributed, common and easily collected to facilitate comparison between sites separated by large distances.
3. *Robust* – bio-indicators should not be killed by very low levels of pollutants (with the exception of type 1 community structure monitoring where sensitivity is important) and should be robust enough to be caged in polluted field sites.
4. *Responsive* – the organisms should exhibit measurable responses to pollutant exposure by having greater concentrations of the contaminant(s) in the tissues (type 2 biomonitoring), by exhibiting effects such as reductions in scope for growth and fecundity, increased incidence of disease or induction of a biochemical response (type 3 biomonitoring), or by possession of genetically-based resistance (type 4 biomonitoring).
5. *Reproducible* – the species chosen should produce similar responses to the same levels of pollutant exposure in different sites.

Further Reading

BAKER A. J. M. and WALKER P.L (1989) Concise review of metal tolerance in plants (see also Ernst *et al.*, 1992).

BRITISH ECOLOGICAL SOCIETY (1990) A useful summary of the background to monitoring water quality.

BRYAN G. W. *et al.* (1986) A classic paper on the effects of TBT on dog whelks.

CALLAHAN C. A. *et al.* (1991) Excellent paper describing *in situ* monitoring with earthworms at a 'superfund' site in the USA.

HOVE K., *et al.* (1990) Interesting study on radiocaesium in goats which emphasizes the importance of long-term monitoring.

LOWE V. P. W. (1991) Use of laboratory data to interpret 'unexplained' mass mortality of seabirds near the Sellafield nuclear reprocessing plant.

NAYLOR C. *et al.* (1989) Study on Scope for Growth in *Gammarus pulex*.

POSTHUMA L. and VAN STRAALEN N. M. (1993) Comprehensive review of metal tolerance and resistance in terrestrial invertebrates.
WRIGHT J. F. *et al*. (1993) A good summary of the RIVPACS technique by some of the workers who developed it.

PART THREE

Effects of Pollutants on Populations and Communities

12

Changes in Numbers: Population Dynamics

So far, this book has described the effects of pollutants on individual organisms, and Chapter 11 described methods of collecting data on populations in the field. In this chapter we consider the uses to which population data may be put, how they should be analyzed, and the interpretations that may be made as to cause and effect. The first five sections of the chapter are an exposition of population ecology theory, and some readers will find this quite heavy going. An attractive route for those encountering this material for the first time may be to begin by reading the case studies in section 12.6.

When pollutants enter an ecosystem, the species within it may be affected in any one of the ways shown in Figure 12.1. The numbers of some species will decline, perhaps to zero (Figure 12.1, curve i) if the species becomes locally extinct. Alternatively, numbers may decline but level out lower than before (curve ii) and the population may persist at this level if the pollution endures

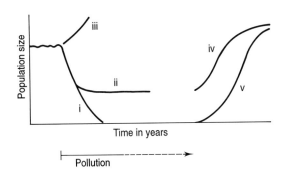

Figure 12.1   Possible responses of population size to pollution.

(*chronic pollution*). A third possibility is that population size initially increases (curve iii). If the pollution is chronic, resistance may evolve within the population, allowing population numbers eventually to increase to a new equilibrium. The evolution of resistance is considered in Chapter 13. If the pollution is transient then the population may eventually recover, either rising from the level to which it was depressed by pollution (curve iv) or returning through immigration/recolonization if pollution had rendered the population extinct (curve v). In these last two cases the population does not necessarily return to its original level. We shall see examples of some of these curves in the case studies described later in this chapter. To begin with, however, we need a little ecological theory with which to interpret and understand the processes that may operate to produce curves like those in Figure 12.1.

12.1 Population Growth Rate

A key feature of population growth or decline, as shown in Figure 12.1, is the rate at which it occurs. The curve i population, for example, seems to be declining at a constant rate. The curve ii population initially declines, but then steadies, and neither grows nor declines – its population growth rate is zero. The curve iii population initially increases in size – its population growth rate is positive. Thus, population growth rate is positive, zero or negative according to whether the population increases, is stationary, or decreases in size.

Population growth rate is the most important characteristic of the population, and population ecologists spend much time and effort measuring it, and establishing the factors that affect it. Some of these factors are described in section 12.2. Right at the outset, however, it is worth noting that there is one set of factors which is particularly important, and particularly difficult to study in practice. These are density-dependent factors, that is, factors which are affected by population density.

Typically, the effect of density-dependence is that population growth rate reduces as population density increases. In the most straightforward case the population density is then stabilized, at a level characteristic of the environment in which the population exists. This level is referred to as the *carrying capacity of the environment.*

It will be immediately apparent that density-dependence is an unwelcome complication for ecotoxicologists. It means that for a full understanding we need to know not only the way that pollution affects population growth rate, but also the way that density-dependence affects it.

Curve iii in Figure 12.1 represents a population that increases in size when the environment becomes polluted. Let us call this species A. If species A were the only one adversely affected by pollutants (e.g. decreased reproductive success, increased mortality rate), its increase would constitute a paradox. The only way species A can increase is as a result of complex interactions with other species. Perhaps some other species, call it species B, exists which depresses species A. If pollution were to depress species B, the consequence for species A could be beneficial. This could come about in several ways. A common case is that species B is a predator of species A; thus, species A increases because the pollutant removes its predators. For instance, the red spider mite increased in numbers when its predators were killed by pesticides. Interactions between species are examined further in section 12.5.

12.2 Population Growth Rate Depends on the Properties of Individual Organisms

Growth rates of a population can be measured in different ways, as shown in Box 12.1.

Box 12.1 Different ways of assessing population growth

The measure we favour is population growth rate, *r*. Population growth rate is defined as the population increase per unit time, divided by the number of individuals in the population. The definition of population growth rate can be understood mathematically as follows. If population size, $N(t)$, is plotted as a function of time, *t*, as in Figure 12.1, then dN/dt represents mathematically the population increase per unit time, in units of animals per unit time. Population growth rate puts this on a *per capita* (i.e. per animal) basis, by dividing by the number of individuals in the population. Mathematically, $r = 1/N \, dN/dt$. This is measured as animals *per capita* per unit time. Thus if population size is 1000 and is increasing by one animal per year, the population growth rate is 0.001 *per capita* per year.

If *r* is constant, then $N(t) = N(0)e^{rt}$. This shows that exponential population increase occurs if population growth rate is constant.

Another measure that is sometimes used is *net reproductive rate*, usually given the symbol λ, defined as

$$\lambda = e^r \tag{12.1}$$

Conversely,

$$r = \log_e \lambda \tag{12.2}$$

λ is the factor by which the population is multiplied each year. Thus if the population doubles each year, then $\lambda = 2$. For example, if $\lambda = 2$ and initial population size is 10, then in subsequent years population size is 10, 20, 40, 80 ... In this case $r = \log_e 2 = 0.693$. Further properties of λ are discussed in the Appendix.

It is intuitively clear that population growth rate depends on individuals' birth and death rates and on the timing of their breeding attempts. Together, these characterize the individuals' life histories. In general, however, mortality, birth and growth rates may change with age. For a complete description of the life history, therefore, we need a record of *age-specific* growth, birth and death rates. Collectively these are sometimes referred to as the organism's vital rates.

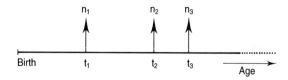

Figure 12.2 A general life history. t_1, t_2, t_3 … represent the ages at which the organism breeds. n_1, n_2, n_3 … are the numbers of offspring then produced by each breeding female.

At this point the reader may wish to look at the real-life example provided in case study 12.1 at the end of this section.

We will here consider two methods of recording a life history that differ in their level of detail. Both can be applied to organisms with discrete breeding events – for example, they might breed annually. We start with the simpler approach.

Suppose the organism breeds for the first time at age t_1, for the second time at age t_2, for the third time at age t_3, and so on, as shown in Figure 12.2. Suppose the number of offspring produced by each female is n_1 at her first breeding attempt, n_2 at her second, and so on. Lastly, suppose that the probability of a female surviving from birth to age t_i is l_i. The population growth rate, designated r, can be calculated from the formula:

$$1 = \tfrac{1}{2}n_1 l_1 l_1 e^{-rt_1} + \tfrac{1}{2}n_2 \, l_2 e^{-rt_2} + \tfrac{1}{2}n_3 \, l_3 \, e^{-rt_3} + \cdots \qquad (12.3)$$

Equation 12.3 is known as the Euler–Lotka equation. In deriving the equation it is assumed that the proportion of females in each age class is invariant (the assumption of 'stable age distribution'). However, estimates of population growth rate calculated from equation 12.3 are still useful in practice even if the stable age distribution assumption does not hold.

It is usual to use a computer to calculate r from measurements of the life-history parameters (the values of n_i, l_i and t_i). The simplest method is to calculate the right-hand side of equation 12.3 for each of a number of trial values of r (e.g. try -0.5, -0.4, -0.3, … 0.3, 0.4, 0.5). Just one value of r makes the right-hand side of equation 12.3 equal to 1. That value of r satisfies equation 12.3 and so measures the population's growth rate.

The life-history may be simpler than that shown in Figure 12.2. For example, adult mortality rate may be constant, or birth rate may be constant, or breeding attempts may be at regular intervals. Such simplifications often allow equation 12.3 to be written in a simpler, more tractable form (Sibly and Antonovics, 1992).

One other very important way of describing life histories is to record the number of organisms surviving and the number of offspring produced at regular intervals, for example daily. These records are conveniently tabulated in a matrix, referred to as a population projection matrix. Powerful methods of matrix algebra have been developed and applied to the analysis of these matrices (Caswell, 1989). A brief introduction to their use is given in the Appendix.

In the following case study, life histories were recorded by daily counting of the numbers of animals surviving, and of the numbers of offspring produced.

Case Study 12.1 The Life History and Population Growth Rate of the Coastal Copepod *Eurytemora affinis* at Different Concentrations of Dieldrin, studied by Daniels and Allan (1981)

A population of *E. affinis* was obtained from Chesapeake Bay, Maryland, where it undergoes annual population expansions between February and May, when water temperatures are increasing from 5 to 15–20°C. Animals were kept in the laboratory for 2 months (three or four generations) at 18°C before experimentation began.

The experiment consisted in recording the life histories of animals subjected to different concentrations of dieldrin. There were seven treatments, 0, 1, 2, 3, 4, 5 and 10 $\mu g \times l^{-1}$ dieldrin, together with an acetone control because acetone was used as the 'carrier' of dieldrin. Sixty newly-hatched larvae ('nauplii') were allocated to each treatment and maintained in groups of six in dishes containing 20 ml of bay water. The water was changed every other day when the animals were fed a fixed number of algal cells. The numbers of survivors and births were counted daily.

The survivorship curves of animals undergoing the various treatments are shown in Figure 12.3A. Survival to day 20 was worse at concentrations of 5–10 $\mu g \times l^{-1}$ than at lower concentrations. Reproduction began around day 18, and birth rate was rather variable over time (two representative birth-rate curves are shown in Fig. 12.3B).

The life history can be fully described by the three parameters: mortality rate, birth rate, and age at first reproduction, if the simplifying assumptions are made that, for each treatment, mortality rate does not change with age and that birth rate does not change with age after first reproduction. Mortality rate, birth rate, and age at first reproduction ('development period'), are shown in relation to dieldrin concentration in Figure 12.4A. As dieldrin concentration increased, mortality rate and development period increased, and birth rate fell.

Any of these effects on its own would produce a reduction in population growth rate. The reductions in population growth rate that would be produced by each effect on its own are shown in Figure 12.4B. The reduction due to birth rate depression is similar both to that due to increased mortality, and to that due to increased development period. Thus, the reduction in population growth rate with increasing dieldrin concentration is due equally to birth rate, death rate, and age-at-first-reproduction effects (Figure 12.4B).

This case study shows how population growth rate varied with dieldrin concentration (Figure 12.4B). Population growth rate also depends on many physical aspects of the environment. For example, population growth rate of

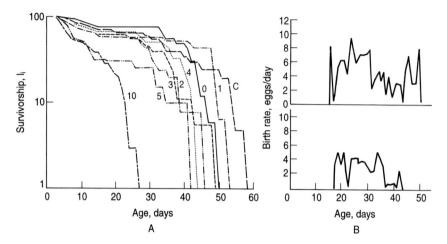

Figure 12.3 Effect of dieldrin on the life history of *Eurytemora affinis*. (A) Survivorship curves. These show for each treatment the survivorship of animals to each age. Numbers indicate treatments, i.e. concentrations of dieldrin, in $\mu g \, l^{-1}$. C is the acetone control. (B) Birth rate in relation to age. Birth rate was measured at all concentrations, but only two representative concentrations, 2 and 4 $\mu g \, l^{-1}$, are shown here (upper and lower graph respectively). Modified from Daniels and Allan (1981).

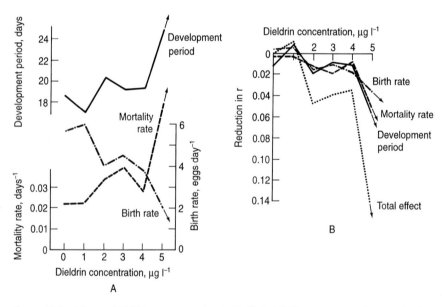

Figure 12.4 Effects of dieldrin concentration in *E. affinis*. (A) Effects on mortality rate, birth rate, and development period (age at first reproduction), estimated as described in the text. (B) Reductions in population growth rate, *r*, caused by the effects shown in (A). From Sibly (1996).

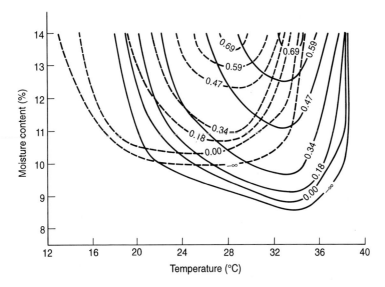

Figure 12.5 Contour plots of population growth rate for two species of grain beetle, *Sitophilus oryzae* (– – –) and *Rhizopertha dominica* (———). From Andrewartha and Birch (1954). (Contours are here labelled in terms of population growth rate; net reproductive rate was used in the original.)

the rice weevil *Sitophilus oryzae* depends on the temperature and moisture content of the grain stores in which it lives. The dependence is shown in a contour plot in Figure 12.5 (dashed lines). The axes represent the temperature and moisture content of the grain. The dashed lines are contours of equal population growth rate. Within the range of environments represented in Figure 12.5 there are some in which the population flourishes (high values of *r*). In general, the higher the value of *r* the faster the population grows. Ecologists refer to those conditions for which $r \geq 0$ as the species' *ecological niche*.

In general, different species have different niches. The grain beetle *Rhizopertha dominica*, for example, flourishes at higher temperatures than *S. oryzae* (Figure 12.5).

In Figure 12.5 population growth rates were not affected by interactions with other animals. When interactions within and between species are taken into account, population growth rates are reduced. In the long term, population growth rates do not exceed zero (i.e. long-term population explosions do not occur).

As we have seen, many factors affect population growth rate. If the long-term population growth rate is zero, however, some factors must operate more strongly when the population is large, but only weakly when the population is small (*density-dependence*). We turn to these factors next.

219

12.3 Density-dependence

As mentioned earlier, factors which vary in their effect with population density are said to be density-dependent. Their net result is that population growth rate is affected by population density. In the simplest case, population growth rate is a negative linear function of population density, as in Figure 12.6. Note that when the population is small (left-hand side of Figure 12.6), the population increases (population growth rate is positive). When the population is large (right-hand side of Figure 12.6), the population decreases (population growth rate is negative). In between, there is an (equilibrium) population density for which population growth rate is zero. This population density is called the *carrying capacity* of the environment in which the population lives. Ecologists give it the symbol K. The effect of density-dependence is here to push the population density towards the equilibrium density, if other factors have increased or decreased it. The equilibrium is therefore a *stable equilibrium*.

Suppose population growth rate reduces with population density in a straight-line relationship, as in Figure 12.6. Let the equation of the straight line be:

$$\text{population growth rate} = r_0 - bN \tag{12.4}$$

where r_0 and b are constants. r_0 represents population growth rate at low population density. We can write b in terms of r_0 and carrying capacity, K, as follows. When the population is at the carrying capacity of the environment,

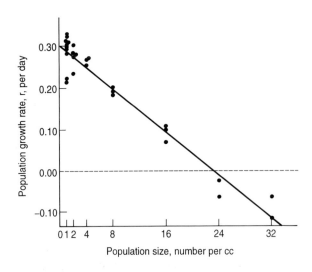

Figure 12.6 Density-dependence in *Daphnia pulex*. From Frank *et al.* (1957).

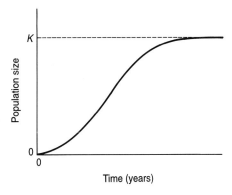

Figure 12.7 The 'sigmoidal' growth curve that results from the logistic equation (12.7). K is the carrying capacity of the environment.

population density is K, and population growth rate is zero. Substituting these values in equation 12.4 we obtain:

$$0 = r_0 - bK \tag{12.5}$$

Hence $b = r_0/K$, and substituting this value of b into equation 12.4 we get:

$$\text{population growth rate} = r_0 - r_0 N/K \tag{12.6}$$

Since population growth rate $= 1/N \ \mathrm{d}N/\mathrm{d}t$ (from Box 12.1), equation 12.6 can be written:

$$\frac{\mathrm{d}N}{\mathrm{d}t} = r_0 N\left(1 - \frac{N}{K}\right) \tag{12.7}$$

Ecologists call this the *logistic equation*. It produces a 'sigmoidal' pattern of population growth (Figure 12.7). When small, the population grows exponentially (left-hand side of Figure 12.7). As population density approaches carrying capacity, population growth rate declines, resulting in a slow approach to the final, equilibrium value (right-hand side of Figure 12.7).

12.4 Identifying Which Factors are Density-Dependent: k-Value Analysis

Population growth rate depends on the life-history traits of the individuals in the population, as described in section 12.2. In particular, population growth rate depends on individuals' age-specific birth and death rates, and on the timing of their breeding attempts. Any or all of these may be density-dependent. The analysis of which traits are density-dependent is referred to as *k*-value analysis by population ecologists. This is because age-specific mortalities are known as *k*-values. Mortality at age (or stage) *i* is given the symbol k_i. In practice it is usually mortalities that are analyzed.

k-Value analysis assesses how the *k*-values vary as population density varies. Usually, natural variation in population density is used. Population density is measured repeatedly, over a period of years, together with the *k*-values. Mortality at age *i*, k_i, is then plotted against population density. The most appropriate measure of population density is usually that of individuals of age *i*. An example of a *k*-value analysis is shown in Figure 12.9. It was obtained from a 25-year study of sea trout by J. M. Elliott and collaborators (Elliott, 1993). By electric fishing at fixed times of year, population density was established at various points in the life history, as shown in Figure 12.8.

k-values, which are measures of mortality, were calculated using the formula:

$$k_i = \log_e R_{i-1}/R_i \tag{12.8}$$

where R_i represents the population density of stage *i*, as indicated in Figure 12.8. These age-specific mortalities are plotted against population density for five phases of the sea trout life history in Figure 12.9. In the first two phases (Figure 12.9, A and B) there are clear positive relationships between population density and age-specific mortality, but there is no relationship in the later phases of the life history. The effect of a positive relationship as shown in Figure 12.9, A and B, is to stabilize the population because higher mortality occurs at higher population density, making the population decrease when population density is high. Conversely, at low population density mortality rate is relatively low, and this allows the population to increase.

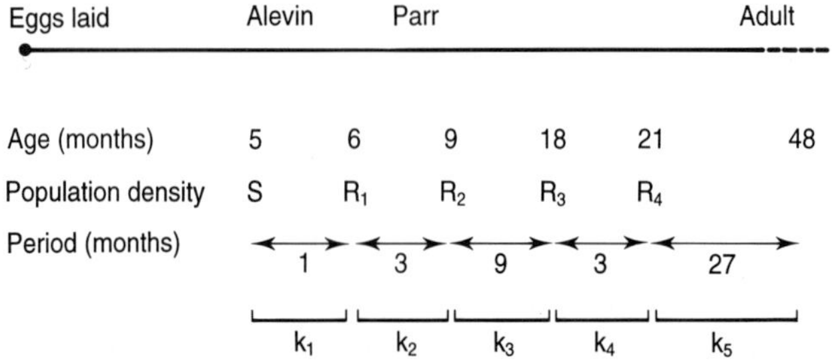

Figure 12.8 The life history of the sea trout at a stream in Northwest England. The eggs hatch after about 5 months. The young trout are known as alevin from hatching until they resorb their yolk sacs; after this they are known as parr. Population density at each age was measured by electric fishing and was designated S, R_1, R_2 ... as shown.

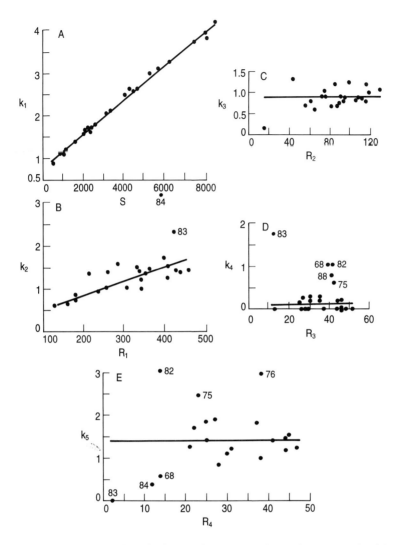

Figure 12.9 Sea trout mortalities, k_1–k_5, in relation to population density in each of the five periods depicted in Figure 12.8. Thus (A) refers to alevin, (B) to young parr, and so on. periods shown in Figure 12.8 using equation 12.8. All data were collected at the same site. Numbers refer to years. From Elliott (1993).

Although population ecologists generally work with k_i-values, as in equation 12.8, it is sometimes better to use mortality *rates*. These can be calculated as:

$$\text{mortality rate} = k_i/t_i \tag{12.9}$$

$$= 1/t_i \log_e R_{i-1}/R_i \tag{12.10}$$

This example shows how population density affects mortality rate, and it may also affect somatic growth rate and birth rate. Moreover, as noted before, when considering the effects of pollutants, mortality, somatic growth and birth rates together determine population growth rate.

The effects of population density on mortality rate are central to population ecology and have been reviewed by Sinclair (1989). Considering the importance of the topic and the attention it has received over the years it is perhaps disappointing to record Sinclair's conclusion that 'we still have a poor understanding of where density dependence occurs in the life cycle of almost every group of animals'.

12.5 Interactions Between Species

So far we have considered some of the factors which affect the numbers of a single population. If understanding a single population is difficult, untangling interactions between species is more than twice as hard. Here population growth rate depends not only on the population's own density, but also on the population density of the other species. To establish these dependencies in the field in generally prohibitively expensive. For this reason detailed study has usually been restricted to simple laboratory systems or to mathematical models.

Despite these difficulties a number of general points can be made.

Interactions between species can be logically classified as being of one of three types:

Competition, in which the population growth rate of each species decreases the more there are of the other species. Generally, species compete for common resources (e.g. food, space, breeding sites), so the more competitors there are, the lower the average success of each.

Mutualism, in which the population growth rate of each species increases the more there are of the other species. Thus, under mutualism, the species in effect help each other. Mutualism is the opposite of competition.

Predator–prey, in which the population growth rate of species A increases the more there are of species B, but the population growth rate of species B decreases the more there are of species A. These population properties are also evident in host–parasite, host–parasitoid and plant–herbivore interactions, and these can therefore be treated under the same heading as predator–prey.

Although, logically, all interactions must fit into one of the above categories, because population growth rates are density-dependent, the interaction of two species may not be in the same category at all population densities.

A species whose numbers vary substantially with time can persist in its environment only if it has a *refuge* that protects it and allows it to increase when its numbers have been reduced to low levels. The term 'refuge' is here used in a very broad sense. It may refer to a physical refuge. Examples of physical refuges include defendable holes or cracks in rocks, as used by whelks to escape predation by crabs, or areas accessible to only one species, such as the splash zones on upper shores where barnacles live but their whelk predators cannot follow them.

Environmental patchiness can also result in refuges for prey species. For instance, examples are known in which infested patches go extinct, but prey are able to escape to uninfested patches. If there is a sufficient time lag before the predators find the uninfested patches, this effectively creates a 'refuge' for the prey in which, for a time, they can increase.

Refuges can occur as a result of predators' foraging behaviour. There are many cases known in which foraging effort decreases as the density of prey decreases, and this may result in a reduced mortality rate of prey at low prey densities, allowing prey populations to grow. Reduced foraging effort at low prey densities may result if predators switch to a different type of prey.

It follows from the definition of predator–prey interactions that removal of predators can lead to an increase in the population density of the prey species. Many cases are known in which pollutants are known to have had these two related effects (Dempster, 1975). For example, the red spider mite, *Panonychus ulmi*, appeared as a pest on outdoor fruit trees after the elimination of the slow-breeding predatory insects which previously controlled the mites. Fruit farmers used to apply pesticides to orchards in Britain as many as 20 times in a season, and this killed the mite's predators, so upsetting the natural balance which had kept the mite population under control (Mellanby, 1967). More recently, Inoue *et al.* (1986), investigating the effects of spraying Kanzawa spider mites, *Tetranychus kanzawai*, with six kinds of insecticide and three kinds of fungicide, showed that the population density of the Kanzawa spider mites increased with the application of certain insecticides and fungicides, probably because of their adverse effects on the natural predators (three species of predatory mite, three species of predatory insect and a spider).

12.6   Field Studies: Four Case Studies

Ecological studies of the effects of pollutants are generally based on circumstantial evidence, though experimental studies are possible (see below). Commonly the time course of pollution is monitored, and compared with observed changes in species numbers. Negative correlations between pollution and species numbers do not, however, necessarily imply causal links, and there is always a worry that parallel changes could have been coincidental. This makes it attractive, as in all science, to carry out replicated experiments in which a treatment is applied to experimental areas, but not to control areas. The use of proper replication allows an assessment of whether the variation between treated and untreated areas is significantly greater than natural variation between control areas. The effect of the treatment can then be measured as the difference between the two types of areas, and this can be assessed statistically provided that a suitable experimental design has been used.

Case Study 12.2 The Decline of the Partridge was Due in Part to the *Indirect Effects of Pesticides Killing the Insect Prey necessary for Chick Survival*

Eighteen years of intensive study of the partridge (*Perdix perdix*) in Sussex, England, are summarized, and worldwide trends are reviewed in a book by G. R. Potts (1986), from which the following account is taken.

The decline in numbers in Britain and worldwide are shown in Figure 12.10. In seeking the reasons for the decline, it is sensible to start by examining the trends in *k*-values. Table 12.1 summarizes data from 34 populations, indi-

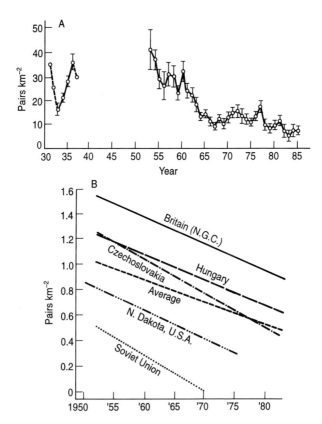

Figure 12.10   (A) The UK Game Conservancy's National Game Census March pair counts for the partridge from 1933 to 1985, with estimated minimum densities for the early 1930s, ± 2 standard errors. (B) The trend in density of breeding pairs km^{-2} over the period 1952–1985 in various regions of the world range. From Potts (1986).

Table 12.1 Comparison of mortality rates in stable and declining populations $\pm$ standard errors[a,b]

		Populations		
		Stable (21)	Declining (13)	Significance
Nest loss	$(k_1 + k_2)$	0.26 ± 0.02	0.21 ± 0.03	ns
Chick mortality	k_3	0.29 ± 0.02	0.44 ± 0.02	$P < 0.01$
Shooting mortality	k_4	0.07 ± 0.01	0.08 ± 0.02	ns
Winter loss	k_5	0.38 ± 0.03	0.41 ± 0.05	ns
Total loss		1.00	1.14	

[a] The last column gives the results of statistical tests comparing stable and declining populations.
[b] From Potts (1986).

vidually studied for between 2 and 29 years between 1771 and 1985. This suggests that the main reason for the population decline lies in the increase in chick mortality, k_3. This raises the question as to what factors cause increases in chick mortality rate?

The causes of partridge chick mortality have long been a matter of controversy among gamekeepers and ecologists. Both weather and the availability of insect food could be important. The advocates of weather are struck by the fact that small partridge chicks can produce only about a third of their own body heat, the rest comes from brooding parents. In cold weather, chicks therefore cannot afford to spend too long away from their parents, so feeding time is limited. This might lead to chick starvation. Arguing on this basis, there have been many attempts to correlate chick production with summer weather, but none of these have been particularly successful.

The other school of thought holds that insects must be an important food for chicks because chicks go to a great deal of trouble to find insects, and eat them in large quantities. Laboratory studies have shown that insects are nutritionally necessary for growing chicks; also in the Sussex study there was a good relationship between chick mortality k_3 and the density of 'preferred insects' (Figure 12.11). Insects were considered to be 'preferred' if they were more than twice as frequent in faecal samples as in vacuum samples from the same fields.

Separation of the effects of insect availability and weather was achieved by multiple regression. This showed that 48% of the variation in chick mortality was explained by the density of preferred insects, and an additional 10% by average daily temperature in the critical period 10 June–10 July. Both of these were statistically significant ($P < 0.001$), but it seems that insect density was much more important than temperature in the Sussex study.

Given that reduced availability of insects is the key to the partridge population decline, it is natural to ask whether the insects declined as a result of

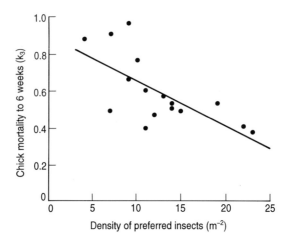

Figure 12.11   Annual chick mortality in relation to the density of preferred insects in the third week of June. Data from the Sussex study 1969–1985. From Potts (1986).

pesticide usage. The increase in the use of herbicides is shown in Figure 12.12, and chick mortality rates k_3 are shown in relation to pesticide usage in Table 12.2. Note that chick mortality rates were considerably increased by the use of herbicides, and increased again when some insecticides were also employed. The effect of herbicides was presumably indirect, removing the food necessary for the survival of the insects eaten by the partridge chicks. The effects of insecticides on insects appear less important. It seems reasonable to conclude

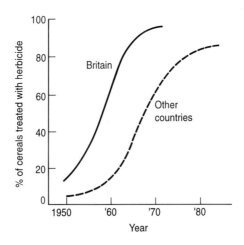

Figure 12.12   Trend in herbicide use on cereals. From Potts (1986).

Table 12.2 Summary of estimates of partridge chick mortality rates grouped according to herbicide and insecticide use[a]

	Mean $k_3 \pm$ SE		
	Up to 1952 (no herbicide)	1953–1961 (some herbicide)	1962–1985 (herbicide + some insecticide)
National Game Census	0.33 ± 0.02	0.45 ± 0.05	0.51 ± 0.02
Damerham	0.32 ± 0.05	0.50 ± 0.06	0.50 ± 0.05
Lee Farm	—	0.44 ± 0.06	0.67 ± 0.07
North Farm	—	0.45 ± 0.05	0.63 ± 0.04
Sussex study	—	0.40 ± 0.05	0.61 ± 0.04
Mainland Europe	0.29 ± 0.03	0.37	0.45 ± 0.03
North America	0.25 ± 0.07	0.28	0.36 ± 0.04

[a] From Potts (1986).

that the decline of the partridge population occurred largely because of increased chick mortality, and that this in turn was the result of a decrease in the density of insects consequent on increased use of pesticides.

The above are not of course the only factors known to affect partridge populations. Nest losses are known to be density dependent, the form of the relationship depending on whether gamekeepers are present (Figure 12.13). Dispersal and adult mortality due to shooting are also density-dependent.

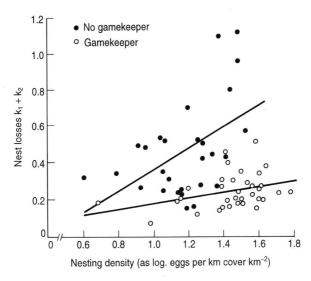

Figure 12.13 Dependence of nest losses on nesting density, with and without gamekeepers. From Potts (1986).

These density-dependent effects would return populations to carrying capacity within a few years were it not for the high levels of chick mortality. Not surprisingly, equilibrium levels are considerably higher if gamekeepers are present.

Because fewer gamekeepers are employed in the UK now than formerly, it is reasonable to ask whether the population decline could better be ascribed to reduced gamekeepers than to increased pesticide use. The Sussex study attempted to answer this by entering the key known relationships into a simple model of partridge population dynamics. The model comprised four basic equations showing how the four k values listed in Table 12.1 were affected by population density and some environmental features. The model was used to see how the population would have reacted if gamekeepers had been employed and if pesticides had not been used. Figure 12.14 shows that, according to the model, use of gamekeepers would have reduced the rate of population decline, but would not have prevented it. Only when herbicides are not used, restoring chick mortality rates to their former levels, is the population decline prevented altogether.

The final message of Pott's (1986) book was that, as far as could then be seen, density-dependence alone would not be sufficient to save partridge populations from extinction. Partridge preservation could only be achieved by increasing the supply of insects to chicks, so reducing the mortality rate of chicks. At the time of writing (1995) little has changed except very locally where appropriate management has been introduced, and the partridge con-

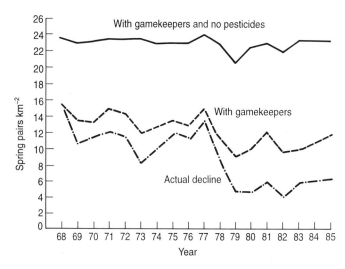

Figure 12.14 Simulations of the Sussex partridge population, showing the actual decline, the less severe decline that would have occurred if gamekeepers had been employed at 1968 levels, and how no decline would have occurred if gamekeepers had been employed and pesticides had not been used. From Potts (1986).

tinues to decline (Potts, in press). This huge loss of a valuable natural resource is arguably worth over £500 million annually in Europe alone.

Case Study 12.3   Population Studies of Pesticides and Birds of Prey in the UK

The study of the effects of pesticides on birds of prey in the UK has a special place in ecotoxicology. This is partly because some pesticide effects were reported first in these species, and partly because of the intensive nature of the studies, which have been conducted over 50 years. Despite this, our knowledge of some of the population processes is less complete than in the case of the partridge. It is therefore necessary to keep in mind that even where pesticides are known to affect individual vital rates (e.g. mortality rate or breeding success) this does not necessarily lead to population decline. Density-dependence can compensate for the effects of the pesticides on particular age classes, and the net result could be that population growth rate is unchanged.

The Peregrine falcon, *Falco peregrinus*, was the first bird of prey to be analyzed in detail from the point of view of pesticide impact. Very similar, and in some instances fuller, data have been obtained for another bird of prey, the Sparrowhawk, *Accipiter nisus*, by I. Newton. Some of these data will be referred to where appropriate. The following account is based mostly on books by Ratcliffe (1993) and Newton (1986). We begin by reviewing the properties of the pesticides involved and their known effects on mortality, breeding success and behaviour, and then go on to consider their effects on populations.

The pesticides that have been shown to affect the birds of prey are insecticides belonging to the organochlorine group. They include DDT, introduced into agricultural use in the late 1940s, and the cyclodiene insecticides aldrin, dieldrin and heptachlor, introduced in the mid 1950s. Both types are extremely stable in their original form and/or as metabolites and so persist in the environment for many years. In addition they are readily dispersed in wind and water or in the bodies of migratory birds and insects.

The organochlorine pesticides have properties which make them particularly hazardous to birds of prey and their predators. Because of their high fat solubility and resistance to metabolism, they have long biological half lives in many species (see Chapter 5). Consequently they tend to bioaccumulate as they move along food chains, reaching their highest concentrations in predators. It follows that top carnivors are especially useful indicators of the environmental effects of these compounds.

In the case of both sparrowhawks and peregrines, organochlorine insecticides have two distinct types of effect upon wild populations. First, the cyclodiene insecticides aldrin, dieldrin and heptachlor (used for example as seed dressing chemicals) can cause lethal toxicity – and so may increase mortality rates. Some sparrowhawks and peregrines found dead in the field contained

231

lethal concentrations of these insecticides in tissues such as liver, brain or muscle. Evidence from both laboratory and field studies indicates that tissue concentration of dieldrin and heptachlor epoxide above 10 p.p.m. will cause death of predatory birds due to their direct neurotoxic action. In some cases, individuals found to contain lethal levels had shown typical symptoms of cyclodiene poisoning before death (e.g. tonic convulsions). These compounds affect the central nervous system by inhibiting GABA receptors. It is known from laboratory studies that they can have sublethal effects on the function of the nervous system (e.g. irritability, convulsions) before lethal effects and pro-duced. There can be no doubt that sublethal effects as well as lethal ones were produced in the field during and after the 1950s. However, there is no means in retrospect of quantifying this. The seriousness of sublethal effects upon the nervous system of predators like the sparrowhawk and peregrine scarcely requires emphasis. A high level of skill and coordination is required in pred-ators if they are to catch their prey. Disturbances of the nervous system can seriously affect their hunting skills (Chapter 8, section 8.4.2).

A second type of effect is eggshell thinning, caused by *pp′*DDE, a persistent metabolite of *pp′*DDT. The association between eggshell thinning in British peregrines and sparrowhawks and exposure of DDT residues was first noted by Ratcliffe (Figure 12.15). Initially it was not clear whether this was a causal relationship, but dosing predatory birds with *pp′*DDE in the laboratory has established that low levels of the chemical do cause eggshell thinning. It is now

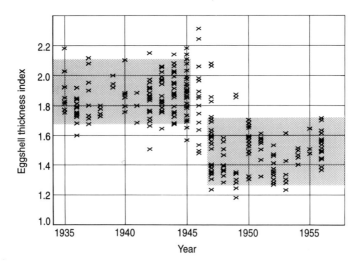

Figure 12.15 The 1947– decline in Peregrine eggshell thickness in the UK. Shaded areas represent 90% confidence limits. Reproduced with permission from *Environmental Reviews*. The eggshell thickness index is defined as: Index = Weight of eggshell (mg)/length × breadth (mm). From Peakall (1993).

known that *pp'*DDE reduces transport of Ca^{2+} to the developing eggshell, probably because of inhibition of calcium ATPase in the shell gland (Chapter 15). Thus, there are good scientific grounds for regarding the close negative correlation between residue of *pp'*DDE and eggshell thickness as a causal relationship (Figure 12.16A). Furthermore, since eggshell thinning is correlated with fledging success in wild sparrowhawks (Figure 12.16B), it is probable that DDT affects hatching success directly. However, although it is well established that lethal toxicity and eggshell thinning have occurred in the field in the UK, difficulties arise in quantifying them, and relating them to population change.

To consider *pp'*DDE and shell thinning first. In both sparrowhawk and peregrine, substantial reductions in shell thickness occurred during 1946–1947 in the UK, coincident with the large-scale introduction of DDT as an insecticide. There was, however, no evidence of a general decline in these species at this time, or for several years after (see Figure 12.17 for peregrine). Since sparrowhawk and peregrine breed at ages 1–2 and 2 years, respectively, and then have relatively short life expectancies (about 1.3 and 3.5 years, respectively), the very sharp decline in populations in the late 1950s cannot be directly attributed to eggshell thinning, which commenced in the period 1946–1947. Thus, although eggshell thinning was occurring, and this was having some effect on fledging success (Figure 12.16B), it did not bring any reduction in population size. Indeed, there was an increase in numbers of peregrines at this time. Presumably, population sizes at this stage were still at or close to carrying capacity, despite the reduction in hatching success brought about by DDT.

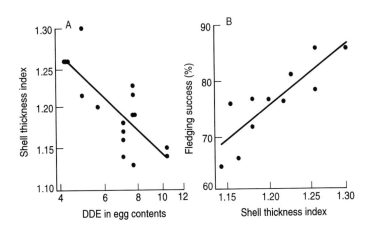

Figure 12.16 In sparrowhawks, (A) DDE is negatively correlated with shell thickness, and (B) eggshell thickness is positively correlated with the number of young raised per brood ('fledging success'). Data from different areas, modified from Newton (1986). Data analogous to (A) for peregrine can be found in Figure 15.4.

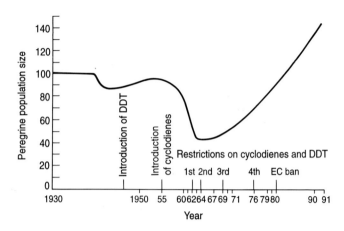

Figure 12.17   Peregrine population size in Britain (1930–1939 = 100) showing the 1961 population decline, and subsequent recovery, together with an outline of pesticide usage. From Ratcliffe (1993).

Probably some density-dependent process compensated for the reduction in fledging success brought about by DDT. Perhaps the smaller number of fledglings were individually more successful in obtaining food, since they had fewer competitors, and so grew faster and were more successful than they otherwise would have been.

Whereas the introduction of DDT did not lead to population decrease, the populations of both species declined rapidly in the mid to late 1950s (see Figure 12.17 for peregrine, Figure 12.18 for sparrowhawk). These declines coincided in both time and space with the introduction of aldrin, dieldrin, and heptachlor as seed dressing chemicals. As discussed above, these compounds caused deaths in the field and extensive sublethal effects were suspected but were not quantifiable.

A series of bans, placed between 1962 and 1975, led to the progressive removal of the cyclodienes and DDT from the UK (Figure 12.17). These bans were followed by a decline in organochlorine residues and population recovery in both of these species. The pattern of recovery differed between areas. Particularly interesting was the late recovery of sparrowhawk populations in Eastern England (Figure 12.18). Sparrowhawk populations recovered when tissue concentration of dieldrin fell below 1 p.p.m. (Figures 12.19, 12.20). Interestingly, the same was true of the Kestrel during this period. Closer inspection of the data for both sparrowhawks and kestrels shows that there was a wide range of tissue dieldrin levels in individuals whose liver samples showed concentrations of 1 p.p.m. or more. Death was attributed to direct dieldrin poisoning (on the grounds of high tissue residues ≥ 10 p.p.m. and symptoms of toxicity) in only 5–10% of the individuals in these samples. However, over 40% of the individ-

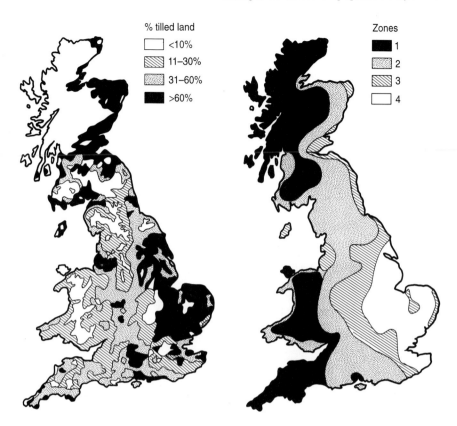

Figure 12.18 Changes in the status of sparrowhawks in relation to agricultural land use and organochlorine use. The agricultural map (left) indicates the proportion of tilled land, where almost all pesticide is used. The sparrowhawk map (right) shows the status of the species in different regions and time periods. Zone 1 – sparrowhawks survived in greatest numbers through the height of the 'organochlorine era' around 1960; population decline judged at less than 50% and recovery effectively complete before 1970. Zone 2 – population decline more marked than in Zone 1, but recovered to more than 50% by 1970. Zone 3 – population decline more marked than in Zone 2, but recovered to more than 50% by 1980. Zone 4 – population almost extinct around 1960, and little or no recovery evident by 1980. In general, population decline was most marked, and recovery latest, in areas with the greatest proportion of tilled land (based on agricultural statistics for 1966). From Newton and Haas (1984), reproduced in Newton (1986).

uals contained residues between 1 and 10 p.p.m. in liver. There is, consequently, a strong suspicion that sublethal effects of dieldrin upon adults were widespread and may have been important in causing population decline; it is also possible that embryotoxicity was a contributing factor.

The recovery of the sparrowhawk shown in Figure 12.20 was rapid between 1980 and 1990, but levelled off in 1993 – a sigmoidal population increase of a

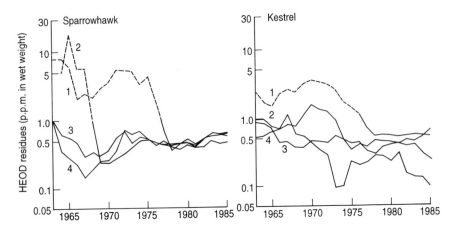

Figure 12.19 HEOD levels in the livers of sparrowhawks found dead in the four zones shown in Figure 12.18. HEOD is the active principle of dieldrin. Broken lines show periods when populations were depleted or decreasing, solid lines show periods when populations were normal or increasing. Population increase occurred when liver levels were less than about 1.0 p.p.m. wet weight. From Newton (1988).

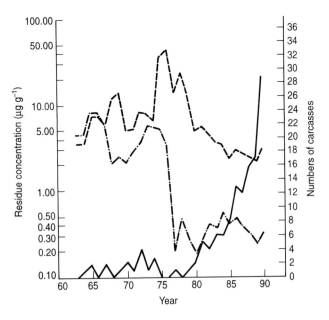

Figure 12.20 Numbers of sparrowhawk carcasses (———) obtained in a region of Eastern Britain, together with concentrations of DDE (———) and HEOD (·—·—·) found in their livers. From Newton and Wyllie (1992).

type frequently encountered as populations approach carrying capacity (compare Figure 12.7). Peregrines also have now recovered (Figure 12.17).

It should be added that the situation was different in North America. Peregrines were less exposed to cyclodienes but more exposed to *pp'*DDE (thus reflecting patterns of use of these insecticides; see also Chapter 15). Eggshell thinning was greater in North American peregrine populations than in British ones. In this case there is strong evidence to suggest that *pp'*DDE was the principal causal agency in population decline, although the field data on populations is not so complete as that obtained in the UK.

The use of eggshell thinning as a biomarker is considered in Chapter 15.

Case Study 12.4 The Boxworth Project. An Experimental Analysis of the Effects of Pesticides on Farmland

The Boxworth project was designed to assess experimentally the effects of three contrasting regimes of pesticide usage on the animals and plants, including crops, on a 300 hectare arable farm in Eastern England. The aim was to investigate large-scale, long-term effects of pesticides under conditions as close as possible to farm conditions. It was a major project, lasting 7 years in its main phase, involving many scientific man-years. The account here is based on Greig-Smith *et al.* (1992). Three regimes of pesticide usage were investigated. These were:

full insurance, in which relatively large amounts of pesticides were applied in advance of possible pest outbreaks, as 'insurance';
supervized, in which pesticides were applied only when needed, as assessed by monitoring pest, weed and disease levels;
integrated, which was similar to supervized, but incorporated some additional features of 'integrated pest management'.

A map of the farm is shown in Figure 12.21. Because large-scale effects were to be investigated, the farm was divided into three large areas, applying one treatment to each area. Because long-term effects were of interest, there was no swapping of treatments between areas.

These features of the experimental design mean that the experiment is unreplicated. For this reason, little assessment can be made of whether the variation between areas is such as might have occurred by chance in the absence of pesticide treatments. The lack of replication precludes the application of the statistical tests normally used in agricultural trials. This could have been remedied by allocating pesticide regimes to fields on, for example, a random basis. Other designs are possible that allow for factors known to cause variation between fields, such as soil type, or the amount of non-crop habitat.

The lack of replication dissipates much of the power of the experimental approach to attribute pesticide cause to ecological effect. However, some idea of natural and baseline variation was obtained by monitoring the areas for 2

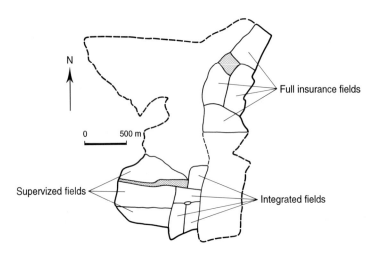

N

0   500 m

Full insurance fields

Supervized fields

Integrated fields

Figure 12.21   Map of the farm at which the Boxworth project was conducted, showing the location of the fields treated with each pesticide regime. From Greig-Smith and Hardy (1992).

years before starting the trials. The pesticide regimes were applied for 5 years (1984–1988), and some residual monitoring has continued since.

In the event there was little difference between the supervized and integrated regimes (S + I henceforth). The analyses therefore generally consist in comparing the effects of full insurance with those of S + I.

Botanical monitoring showed that patterns of grass weed densities reflected the efficiency of the weed control regimes. An example is shown in Figure 12.22A. Full insurance was most effective, and held all grass weeds at low densities. By contrast, the densities of the broad-leaved weeds varied little between regimes (Figure 12.22B). Further botanical comparisons were hampered by high variation between years, including the two baseline years.

Turning to the invertebrates, it appears that, overall, the density of herbivorous invertebrates was about 50% less under full insurance than under S + I. Worst affected were certain non-dispersing species. Carnivorous invertebrates (predators and parasites) showed a similar pattern, but detritivores were unaffected.

It proved particularly difficult to ascribe cause and effect in small mammals and birds, partly because of their mobility. The diet of common shrews (*Sorex araneus*) reflected the distribution of the invertebrates, described above, that form their diet. In particular, leatherjackets (crane-fly *Tipulidae* larvae) were less frequent in the stomachs of shrews caught in the full insurance area than in the supervized area. However there was no evidence of long-term effects of the different pesticide regimes on any of the small mammal populations studied. Autumn application of slug pellets containing methiocarb had an immediate local impact on adult woodmice (*Apodemus sylvaticus*), but the

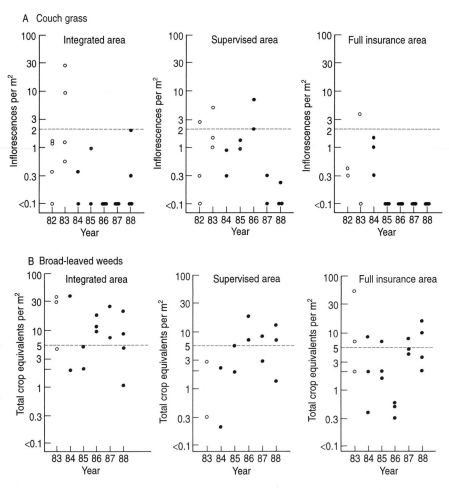

Figure 12.22 Efficacy of the Boxworth pesticide regimes on the densities per m² of (A) the weed grass Couch (*Elymus repens*) in July, and (B) the broad-leaved weeds in spring. 1982 and 1983 were 'baseline' years before the pesticide regimes were applied. The dotted lines show 'spray decision thresholds' used in deciding whether to apply pesticides. From Marshall (1992).

population recovered quickly by immigration of juveniles from nearby woods, fields and hedges.

The only bird species affected by the full insurance regime was the starling (*Sturnus vulgaris*). Part of the observed reduction in numbers of starlings nesting in the full insurance area relative to the S + I area may have been due to the effects of the pesticides on leatherjackets.

The economics of the pesticide regimes were also evaluated. Yields of wheat in the full insurance fields were 0.92 tonnes per hectare higher than in the supervized fields, and 1.35 tonnes per hectare higher than in the integrated area, although there was considerable variation from year to year. Grain quality also appeared better under full insurance. However the extra costs inherent in full insurance meant that the supervized approach was as profitable as full insurance. The integrated regime used in the study was less profitable. However truly integrated low-input systems might do better.

The Boxworth project shows clearly that the experimental study of pesticide effects in the field is a very expensive business. A replicated experimental design would have allowed stronger conclusions to be drawn, but since the farm would have been divided into smaller experimental plots, mobility between areas would have increased, and mobility was already a significant factor affecting the distributions of small mammals and birds. The lack of replication means that the Boxworth project was essentially a pilot study. It does nevertheless suggest which forms of wildlife were, and which were not, affected by high rates of application of pesticides.

Further Reading

BEGON, M. MORTIMER AND THOMPSON, D. J. (1996) *Population Ecology: A unified Study of Animals and Plants* Provides an introduction to population ecology, including the analysis of density-dependence and interactions between species.

EVANS, P. R. (1990) Provides a useful discussion of the population effects of pesticides on birds and mammals.

13

Evolution of Resistance to Pollution

Evolutionary responses to pollution are referred to as resistance. It is implicit that resistance has a genetic basis, and what is known of its genetic inheritance is briefly outlined in section 13.5. This chapter considers how genetic changes in resistance come about. Resistance represents an evolutionary response to environmental changes resulting from pollution, and the first sections of this chapter describe the general phenomenon of evolutionary response to environmental change. Pollution represents an environmental change for the worse, and resistance generally defends organisms against the deleterious consequences of pollution. Such defence may reduce an organism's mortality rate, but this is sometimes expensive, using energy and/or nutrients that could otherwise have been used for reproduction or somatic growth. Defence may, therefore, involve a trade-off between production and survival: increased survival may only be obtained at a cost of reduced growth or reproduction. As a result resistance may have a fitness cost in an unpolluted environment. The possible physiological basis of this trade-off has been considered in Chapter 8, the evolutionary implications are considered here in section 13.3. Four case studies of evolutionary responses to pollution at the end of the chapter consider the evolution of pesticide resistance, of heavy-metal tolerance in plants, of industrial melanism, and of TBT resistance in dog whelks. These include some of the best documented studies of evolutionary responses to environmental changes. The first sections of this chapter are quite theoretical, and an attractive approach for those encountering this material for the first time may be to begin by reading the case studies.

13.1 Chronic Pollution is Environmental Change

Transient pollution by definition has only passing effects on gene frequencies. Chronic pollution, however, can have lasting effects, because it changes the environment in which organisms live. In considering evolutionary responses to chronic pollution we are, therefore, dealing with a particular case of the general phenomenon of evolutionary responses to environmental change. Before considering the effects of environmental change, however, we need to know what happens in unchanging environments.

13.2   The Evolutionary Process in a Constant Environment

Consider Figure 13.1, which provides a simple graphical account of the main ideas. The account here is based on Sibly and Antonovics (1992). The evolutionary process is envisaged, put very simply, as consisting of the creation (by mutation) of new alleles, which either displace or are displaced by their counterparts. In Figure 13.1, alleles A–E affect two life-history traits of carriers. These might, for example, be juvenile growth rate and juvenile survival. If the constant environment assumed here is a polluted environment, then alleles C–E might represent resistant alleles that confer improved survival (e.g. allele C) or growth (e.g. allele E). Alleles A and B would then represent non-resistant alleles, whose carriers on average have reduced survival or growth rate. Note that most individuals carry the A allele in Figure 13.1A and so have small values of both traits, but a few individuals have larger values so that overall there is a small positive correlation between individuals. If there is not much effect of environmental variation, this reflects a positive genetic correlation (e.g. the two traits are correlated because they are determined by alleles that have an effect on the magnitude of both traits).

Figure 13.1B represents a hypothetical situation at some later time in the selection process. Now most of the small-trait A alleles have disappeared but the numbers of the large-trait alleles (C–E) have increased. Furthermore, the genetic correlation between individuals is now negative, whereas earlier it was positive.

Clearly, all depends on whether or not an allele spreads in the population – i.e. on the rates of increase of the alleles. The *per capita* (or more correctly *per copy*) rate of increase of an allele is here called its *fitness*. Since the rates of increase of alleles depend on their effects on life histories, the rates of increase (fitnesses) can be plotted out in the space of Figure 13.1, as shown in Figure 13.1C. In our example the small-trait alleles have negative rates of increase, because they are declining in the population. For these alleles fitness is negative (e.g. $r = -1$ in Figure 13.1C). On the other hand, the large-trait alleles have positive rates of increase, since they are spreading. For them fitness is positive (e.g. $r = +1$). Evolutionary change can also occur, however, if both small- and large-trait alleles increase but at different rates.

In general, as selection proceeds, the cloud of points in Figure 13.1 changes shape. In the absence of environmentally-caused variation, the shape of the cloud is measured by genetic correlations and variances, and as the cloud

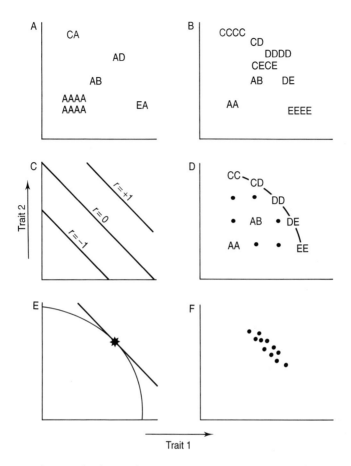

Figure 13.1 Simple example of an evolutionary process. Axes represent two life-history traits. Note that the alleles far from the origin (C–E) have increased in numbers between graph A and graph B, whereas those near the origin (A) have decreased. Graph C shows the per-copy rates of increase, i.e. fitnesses, here labelled *r*. Graph D shows the genetic options set with genotypes obtainable by recombination represented as dots. The boundary of the options set is the trade-off curve (thick line). Graph E shows the optimal strategy (evolutionary outcome), starred. Graph F shows the genetic options that may persist in the population at the end of the evolutionary process. See text for further details.

changes shape, the genetic correlations and variances change accordingly (cf. Figure 13.5). In the absence of further mutation, where would this process end up?

In considering the eventual outcome of this selection process it must be remembered that we are here restricting attention to a constant environment. It is important to realize that the environment of an individual depends not only on physical features (e.g. temperature, rainfall, concentration of a pollutant) and biotic features determined by other species (e.g. food availability, predation) but also has characteristics determined by conspecifics, such as territory size, availability of mates, competition for food, and so on.

In this environment many alleles will affect life-history components. Plotting out all these genetically-codable options (including all possible recombinants) in a space like that of Figure 13.1A gives us a set of points we shall call the *genetic options set* (Figure 13.1D). Although the exposition here is in terms of two-dimensional examples, the concepts all have natural generalizations to three or more dimensions (Sibly and Antonovics, 1992).

Of particular interest, because it limits selection, is the boundary of the options set. We shall call this the *trade-off curve* (Figure 13.1D). This trade-off curve represents the best that this organism can achieve genetically in the study environment.

Putting together the information about fitness (Figure 13.1C) with the information on options sets (Figure 13.1D) the *optimal strategy* is readily identified (Figure 13.1E) as that having the highest fitness in the study environment. This point, then, represents the eventual outcome of selection in this environment.

In this section a distinction has been made between the process and the outcome of selection. The process is modelled by *quantitative genetics*. This determines the short-term evolutionary trajectories within local populations, from knowledge of the fitness surfaces together with the genetic options set, characterized by genetic correlations and variances. The outcome of selection can be identified by ecological optimality theory given knowledge of the shapes of the trade-off curve and the fitness contours. Note that the outcome of selection may be a number of alleles having similar fitness, i.e. a thin oval set of genetic options on or near the trade-off curve in the neighbourhood of the optimal strategy. In the example of Figure 13.1F these genetic options would be characterized by a negative genetic correlation. In this way genetic correlations can provide evidence about the shapes of trade-off curves.

If there are no trade-offs, so that life-history traits can be genetically altered independently of each other, then there is always selection to increase fecundity, decrease mortality rate and breed early. One way to achieve early breeding is through faster growth and development, so a corollary of selecting for early breeding is that there is always selection for faster growth and development – in the absence of trade-offs.

In this section we have seen that the fitness of an allele can be measured by its rate of increase, or more specifically, by its 'per capita' rate of increase. In population genetics it is usual to make the definition of fitness relative to the rate of increase of the most successful allele or genotype, but that approach is not followed here. The advantage of the present approach is that the fitness of an allele can be related directly to the life cycle of its carriers. Fitness is increased by reductions in mortality rate, increases in fecundity, or by breeding earlier. Thus alleles are selected which reduce their carriers' mortality rate, increase their fecundity, or make them breed earlier. It is important to note that the fitness of an allele depends on the environment in which its carriers live. If the environment changes, the fitness of the allele may change.

In these terms an allele can be defined as *resistant* if it increases the fitness of its carriers in a polluted environment. Non-resistant alleles are also known as susceptible alleles. Thus resistant alleles are favoured in polluted environments. What happens in unpolluted environments? If resistant alleles are then outperformed by susceptibles, the resistant alleles are said to have a *fitness cost* . Fitness costs are discussed further in section 13.4.

13.3 The Evolution of Resistance when there is a Mortality/Production Trade-off

It follows from the above that if alleles exist that affect production rate but not mortality rate, selection acts to maximize production rate. Conversely, if alleles affect mortality rate but not production rate, selection acts to minimize mortality rate. However it may not be possible to alter one life-history trait without affecting the other, if mortality and production are involved in a trade-off. A decrease in mortality rate can then only be achieved at the expense of a decrease in production rate. Possible physiological reasons for such a trade-off were described in Chapter 8 in section 8.5.

The action of selection on such a trade-off has been analyzed by Sibly and Calow (1989). They focussed on the juvenile phase, although the analysis can equally well be applied to adults. For the purposes of the analysis they defined 'somatic growth rate' as the reciprocal of development period. This definition makes most biological sense if egg size and adult size are not subject to selection, at least for the period of time analyzed.

For the reasons given in Chapter 8, it is likely that a decrease in mortality rate can only be achieved at the cost of a decrease in somatic growth rate. Protective mechanisms that decrease mortality rate are often thought of as defences. Resources allocated to these defences are not available for somatic growth, and it follows that increasing allocation to defence implies reduced somatic growth. Defensive structures and processes therefore have two key characteristics, from an evolutionary point of view. These characteristics are their effects reducing (1) mortality rate, and (2) growth rate. Plotting out the different genetic possibilities reveals the form of the trade-off curve (Figure 13.2A, cf. Figure 8.14).

It is important to remember that the form of the trade-off curve depends on the organism's environment. What is an effective defence in one environment may have no impact in another – defences against pollution are of no use in unpolluted environments. Thus, trade-offs are environment-dependent.

The fitness contours for this trade-off are straight lines, and the most interesting contour, that on which fitness is zero (stable population), goes through the origin (Figure 13.2B). Superimposing Figures 13.2A and 13.2B reveals the

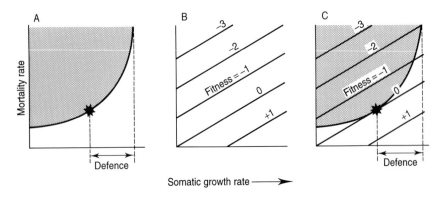

Figure 13.2 (A) Genetic options set (shaded) and trade-off curve. (B) Fitness contours. Note that the zero-fitness contour goes through the origin. (C) Superimposing (A) and (B) allows identification of the evolutionary outcome as the allele achieving highest fitness.

position of the 'optimal strategy' (cf. Figure 13.1E), representing the evolutionary outcome in the studied environment (Figure 13.2C). This is the strategy towards which selection drives the population. It represents the optimal trade-off between mortality rate and growth. It indicates the optimal amount the organism should spend on defence.

This analysis depends, as has been emphasized, on the constancy of the environment. What happens if the environment changes, as when it becomes polluted?

13.4 Evolutionary Responses to Environmental Change

Pollutants may affect the shape and/or the position of the genetic options set. Changes in position will be discussed first. Pollution may damage organisms, either with lethal effect, or with some detriment to production rate. Such processes increase mortality rate, or reduce somatic growth rate, moving the options set either vertically upwards or horizontally to the left.

The outcome of the evolutionary response to such changes in the position but not the shape of the genetic options set is shown in Figure 13.3. The analysis is straightforward. As in the last section, attention can be restricted to steady-state outcomes, in which the final value of fitness is zero. As before, the zero-fitness contours are straight lines going through the origin. Inspection of Figure 13.3 shows that the evolutionary response is less defence if either mortality is increased, moving the options set vertically upwards, or if production is reduced, moving the options set horizontally to the left. In other words either mortality stress or production stress elicits the same evolutionary response – less defence.

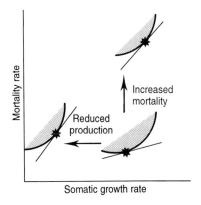

Figure 13.3   The evolutionary outcomes of long-term mortality and production stresses. The straight lines are zero-fitness contours.

This prediction – less defence in polluted environments – appears paradoxical, and it should be emphasized that it applies only if the shape of the trade-off curve remains unchanged when its position is shifted. If more defence evolves in polluted environments then the inference must be that the shape of the trade-off curve has changed. A hypothetical example is shown in Figure 13.4. In this example genes for optimal defence are selected in the polluted environment, and genes for no defence are selected in the unpolluted environment, as shown by the stars in Figure 13.4.

In general, pollution does change the shape of the options set. This may happen because pollution elicits the expression of genes that would not otherwise have been expressed. For instance enzymes may be induced, or rates of pumping or moulting may increase, with consequent effects on life histories. Genetic variation that was of little consequence in the unpolluted environment may now distinguish survivors from non-survivors. Survival may depend on having alleles that increase relatively impermeable exterior membranes (e.g. Oppenoorth, 1985; Little *et al.*, 1989), more frequent moults (and consequent removal of toxicant in shed skin, e.g. Bengtsson *et al.*, 1985), a more comprehensive immune system, or detoxication enzymes (Terriere, 1984; Oppenoorth, 1985), as described in Chapters 4, 5 and 7. Many examples are given in this book.

Figure 13.4 shows that if genes are expressed in polluted environments that before were silent or absent, then evolutionary outcomes may differ between polluted and unpolluted environments, and their populations may be genetically distinct. Such genetic differentiation can be documented and used experimentally in a *transplant experiment*. The method involves transplanting individuals from a number of environments to a common environment in which their life-history components are then measured. Care must be taken

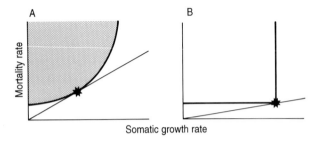

Figure 13.4 If the trade-off curves in (A) polluted and (B) unpolluted environments have different shapes, then the evolutionary outcomes (*) may involve more defence in polluted environments, as shown.

that there are no maternal of other residual non-genetic effects carried over from the previous environment. For instance mothers of measured individuals must be in equivalent condition, since maternal quality can affect offspring performance ('maternal effect'). Since all individuals are assessed in a common environment, population differences must then reflect genetic differences. If the transplanted populations are genetically distinct, it is unlikely that the transplanted populations, carrying alleles that evolved in other environments, will be superior in the study environment to the population that evolved there. Hence, in each environment the population that evolved there should outperform the others. This prediction has often been confirmed by reciprocal transplant experiments (especially in plants) showing that resident populations outperform aliens (Sibly and Antonovics, 1992). Examples comparing resistant and susceptible strains of various animals in polluted and unpolluted environments are given in Table 13.1. The three studies shown all indicate that, as predicted, resistant strains are fitter in polluted environments, and susceptible strains fitter in unpolluted environments. When this happens there is a *fitness cost of resistance*, meaning that resistant alleles, which are fitter in the polluted environment, are less fit than susceptibles in the unpolluted environment. Note, however, that whereas resistant strains are often much fitter than sus-

Table 13.1 Fitness advantage of resistant strains in environments with/without pesticides[a,b]

Species	Pesticide	Fitness advantage with pesticide	Fitness advantage without pesticide
Mosquito (*Anopheles culifacies*)	DDT	0.34	−0.23
Mosquito (*Anopheles culifacies*)	Dieldrin	1.50	−0.49
Rat (*Rattus norvegicus*)	Warfarin	≫0	−0.62

[a] Fitness advantage = Fitness$_{\text{resistants}}$−Fitness$_{\text{susceptibles}}$. Fitness cost is the negative of Fitness advantage.
[b] Data from Bishop (1981) and Curtis *et al.* (1978).

ceptibles in the polluted environments, the fitness of the resistants in the unpolluted environment is sometimes not much less than that of susceptibles (e.g. 0.23–0.62 in Table 13.1). In general, the fitness costs of resistance depend on the resistance mechanism involved. Examples are discussed in the case studies at the end of this chapter. One experimental problem that arises in practice is that if one is interested in the fitness benefits and costs of a single allele, then it is desirable to study the allele and its alternate on a common genetic background.

In general, then, chronic pollution results in a change in the shape of the genetic options set. Whereas before the environmental change the genetic options set may have been lined up on or alongside the trade-off curve and parallel to a fitness contour (Figure 13.1F), after the environmental change, with genes expressed that before were silent, the genetic options set is quite likely to bulge out, and to lie some way from the new trade-off curve. These predictions can be tested by measuring quantitative genetics parameters, which give some idea of the shape of the genetic options set. The degree to which the points are spread out in Figure 13.1 is measured by *additive genetic variance*. The prediction here is that additive genetic variance will increase if new genes are expressed in response to environmental pollution. The correlation between the points in Figure 13.1F is measured by genetic correlation, and since in the unpolluted environment we argued that the genetic correlation would be tight (close to −1 Figure 13.1F), any loosening of the correlation will move it away from ±1.

Holloway *et al.* (1990) tested these predictions by transplanting a population of weevils from the environment in which it had evolved into a new toxin-rich environment. The new, toxin-rich environment consisted of yellow split-pea, and the old environment of wheat, on which substrate the population had been maintained for 50 generations. Measurement of quantitative genetic parameters was achieved in a 'full-sib/half-sib' breeding experiment, using over 50 males and 250 females. This illustrates the general point that quantitative genetic experiments are very demanding in terms of the numbers of animals needed. Because the population declined initially after transfer to the toxin-rich environment, genetic analysis could not be carried out until the fifth generation after transfer. The results are shown in Figure 13.5. Additive genetic variance in development period and mortality rate increased as predicted after introduction to the new, toxin-rich environment. Genetic variance in oviposition rate did not change. Changes in genetic correlation were quite large in two cases (shown in Figure 13.5B). The genetic correlation between oviposition and juvenile mortality rate changed as predicted from a tight correlation close to +1, to a looser correlation of 0.4 in the toxin-rich environment (Figure 13.5B). A correlation of +1, not −1 as in Figure 13.1F, was predicted in the unpolluted environment because although the association between fitness and oviposition rate is positive, that with mortality rate is negative. There was no change in the correlation of mortality and development period.

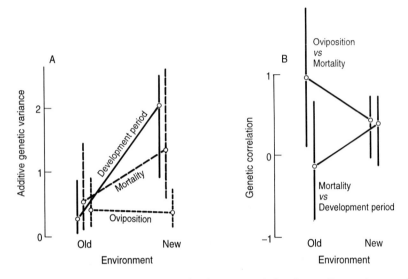

Figure 13.5 Additive genetic variance in development period and mortality rate increased as predicted after introduction of rice weevils *Sitophilus oryzae* to a new, toxin-rich environment, although the changes were not significant. Units are eggs day^{-1} female^{-1} (oviposition rate); 4 days (development period); or 4×10^{-5} (mortality rate). (B) Changes in genetic correlations. Vertical bars represent 95% confidence intervals. From Holloway *et al.* (1990).

This example shows how the genetic options set, measured by quantitative genetic parameters, may change when the environment becomes polluted. If the environment remains chronically polluted, selection will then favour alleles close to the trade-off curve in the neighbourhood of the new optimum. A new evolutionary process begins, but it is a process of the same type as that described in section 13.2.

13.5 Resistance is often Monogenic

Alleles selected in polluted environments, with positive fitnesses, are said to be resistant, or, in some contexts, tolerant. Individuals that are homozygous resistant are referred to as resistant individuals. The relative resistance of heterozygotes measures the degree of dominance of the resistant allele. An example is shown in Figure 13.6. Pesticide resistance often shows simple inheritance, with resistant genes being semidominant (i.e. heterozygotes half way between the homozygotes).

The degree of dominance affects the speed of spread of an allele, but not the final outcome of the evolutionary process, (except when the allele is over-

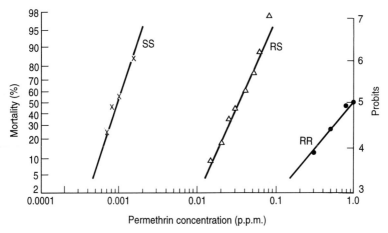

Figure 13.6   Dose–response curves for the mosquito *Culex quinquifasciatus* tested with permethrin (NRDC 167). Percentage mortality is plotted on a probit scale. SS shows the response of homozygous susceptible individuals, RS of heterozygotes, and RR of homozygous resistant individuals. From Taylor (1986).

dominant, i.e. the heterozygotes outperform both homozygotes). Advantageous dominant alleles spread faster initially than recessive alleles.

Complications arise if more than one locus is involved in resistance. To tell how many loci are involved, it is generally necessary to carry out a breeding experiment lasting several generations. For such experiments, homozygous strains are required. These are usually obtained by mass selection in the laboratory of field-collected strains. A breeding experiment starts by crossing a homozygous resistant strain with a homozygous susceptible strain. The offspring are necessarily heterozygous at all loci. These offspring are 'backcrossed' with the parental strains. Half the offspring of these backcrosses are expected to be heterozygous if only one locus is involved. On the other hand, if more than one locus is involved, less than half these offspring show resistance. The exact prediction depends on the degree(s) of dominance of the resistant allele(s). Statistical techniques are available for estimating the number of genes involved, but these make assumptions whose validity has to be checked. When discrimination between genotypes is difficult, as it can be in resistance studies, further experiments may be necessary. Repeated backcrossing can be useful. Genetic markers that map the positions of the resistant genes on the chromosomes have also proved particularly useful in difficult cases.

The general conclusion of studies of this type is that major genes (i.e. genes with large effects) are found in most cases of resistance, including resistance to insecticides, acaricides, fungicides, herbicides and heavy metals. Resistance in not always monogenic, but in most cases where resistance causes problems in pest control, resistance appears to be largely controlled by one, or occasionally two, loci (Roush and Daly, 1990). Minor genes and modifier genes may, however, still have some small effects.

13.6 Case Studies

Examples of evolutionary responses to pollution include some of the best known studies of the evolutionary process in the field. The reason for this is not hard to find. To study the evolutionary process it is usually necessary to find a population exposed to an environmental change, and in recent centuries most environmental changes have been effected by man. Many of these are cases of episodic or chronic pollution.

13.6.1 *Evolution of Pesticide Resistance*

The evolutionary response to pesticides has been staggeringly varied and successful. The number of species in which resistance is known rose from 30 in the 1950s to over 450 in the 1980s. The account here is based on Taylor (1986), Mallet (1989) and especially Roush and Daly (1990).

Most insecticides now in use belong to one of four classes: (1) organochlorines, like DDT, gamma-HCH and dieldrin; (2) organophosphorus compounds (OPs), such as malathion and dimethoate; (3) carbamates, like carbaryl; and (4) synthetic pyrethroids, related to natural pyrethrins, and sythetics such as permethrin. Insects may be resistant to more than one insecticide, and often to insecticides in more than one class. When resistance to more than one insecticide is achieved by a single mechanism, this is true *cross resistance*, but when several resistance mechanisms are involved, this is called *multiple resistance*. In some cases the situation may be uncertain and these terms are then used more loosely.

All members of these four classes of insecticide act on the nervous system, though in different ways. The organochlorines and pyrethroids interfere with electrical conduction along the axons, while the OPs and carbamates act as inhibitors of acetylcholinesterase (Chapter 7).

The mechanisms by which insects have evolved resistance to these chemicals are summarized in Table 13.2. Behavioural mechanisms include avoidance of an insecticide after low level exposure to it. More important in practice are the various types of metabolic resistance, that increase the rate at which the insecticide is broken down. A third type of resistance mechanism involves the animal becoming less sensitive to an insecticide. Often this is due to the site of action being insensitive to the chemical in the resistant strain (Table 13.2). Lastly, the cuticle may be relatively impermeable to insecticides in resistant strains.

All of these resistance mechanisms could involve fitness costs in areas where there is no insecticide. Avoidance of certain microhabitats could result in reduced nutrient uptake, increased detoxication may be energetically expensive, and reduced sensitivity or penetrance of insecticides may involve a costly reduction in sensitivity or penetrance of other substances. Although the fitness costs of resistant alleles in untreated areas have rarely been studied, it is known that they can be large (Table 13.1) Roush and Daly (1990) conclude

Table 13.2 Primary mechanisms of resistance[a]

Mechanisms	Insecticides affected
1. Behavioural	
increased sensitivity to insecticide	DDT
avoidance of treated microhabitats	Many
2. Increased detoxication	
DDT – dehydrochlorinase	DDT
microsomal monooxygenase	Carbamates
	Pyrethroids
	Organophosphorus compounds
glutathione transferase	Organophosphorus compounds
hydrolases, esterases	Organophosphorus compounds
3. Decreased sensitivity of target site	
decreased sensitivity of	
acetylcholinesterase	Organophosphorus compounds
	Carbamates
decreased nerve sensitivity (K_{dr})	DDT
	Pyrethroids
decreased nerve sensitivity	Dieldrin and other cyclodienes γHCH
(GABA receptor)	
4. Decreased cuticular penetration	Most insecticides

[a] Modified from Taylor (1986).

that the most serious and consistent fitness costs are those associated with general esterases (e.g. carboxylesterases that hydrolyze naphthyl acetate). By contrast, in some studies little or no reproductive disadvantage has been found associated with malathion-specific carboxylesterases, increased oxidative detoxication, 'mutant' acetylcholinesterase, and knock-down-resistance-like mechanisms.

Insecticide resistance is characterized by rapid evolution under strong selection. However, resistant alleles do not generally spread to fixation, completely displacing competitor alleles, in the field. This is probably because pesticide treatments are stopped when they become ineffective. However inability to spread to fixation would also result if resistant alleles have a fitness cost in untreated environments. Untreated populations mix with treated populations by dispersal. Immigration from untreated populations can considerably delay the increase of resistance, depending on the relative sizes of the treated and untreated populations, and on the degree of migration.

One of the most spectacular examples of the evolution of insecticide resistance is that shown in Figure 13.7. As part of a resistance management strategy, pyrethroid use was restricted to a short period ('stage II') each year in the middle of the cotton-growing season. Endosulfan use was permitted in stages I and II; other insecticides could be used at any time. Figure 13.7 shows that the

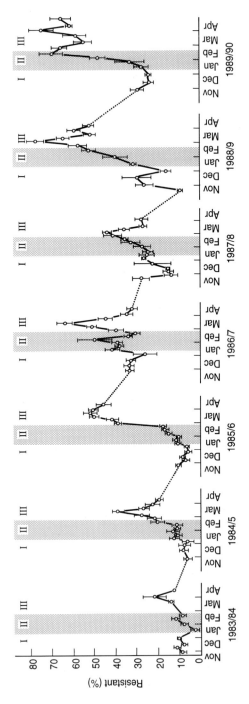

Figure 13.7 The evolution of pyrethroid resistance in cotton budworm, *Helicoverpa armigera* in the Namoi-Gwydir cotton-growing region of New South Wales, Australia. The 'stages' of the resistance management programme are indicated by roman numerals at the top of the graph. The period of pyrethroid use each year ('stage II') is shaded. Percentage resistant refers to the percentage surviving a dose of the pyrethroid fenvalerate that killed 99% of susceptibles. Vertical bars are standard errors. From Forrester *et al.* (1993).

proportion of individuals that were resistant rose each year during and immediately after pyrethroid application. The apparent rise immediately after pyrethroid application ceased is because of a time lag in measurement: affected individuals were not tested themselves, instead the test was carried out on their field-collected eggs. There was thus a delay while individuals matured and reproduced before the test was performed. Ignoring this delay, after pyrethroid application ceased the proportion of resistant individuals declined. This could be the result of a fitness cost of pyrethroid resistance, but it could also be because of immigration of susceptibles from adjoining untreated areas. Note that overall the frequency of pyrethroid resistance rose during the course of the 7-year programme.

13.6.2 *Evolution of Heavy Metal Tolerance in Plants*

The mining of heavy metals inevitably leads to pollution of the soils around mines, and spoil heaps are often rich in copper, zinc, lead or arsenic. The spoil heaps of Devon Great Consols, for example, which was the richest copper mine in Europe in the late nineteenth century, contain more than 1% copper and 5% arsenic in places. Such concentrations are highly toxic to most plants. Yet plants can often be found growing on spoil heaps, as a result of the evolution of metal tolerance (Macnair, 1987; Schat and Bookum, 1992). These plants are genetically more tolerant than plants from neighbouring populations. Methods of quantifying metal tolerance in plants are described in section 6.2.4.

As an example, consider the wind-pollinated grass, *Agrostis tenuis*, copper-tolerant genotypes of which grow on the spoil heaps of copper mines. Taking a transect through such a mine, as in Figure 13.8, one finds copper-tolerance on the copper-rich soils of the mine, but no tolerance outside it in the upwind direction. Some tolerance is, however, found immediately downwind as a

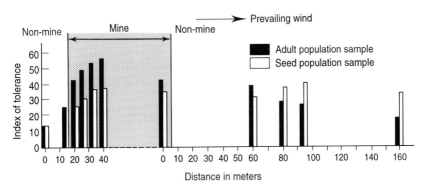

Figure 13.8 Copper tolerance in the grass *Agrostis tenuis* along a transect on the surface of a copper mine. The copper-impregnated part of the mine is shaded. Data of McNeilly (1968) redrawn by Macnair (1981).

result of pollen or seeds of tolerant parents being blown off the mine. Since tolerance is found on the mine but for the most part not off it, it is clear that tolerant alleles have increased in numbers on the mine. This is what is meant by saying that tolerant alleles are fitter in the polluted environment. Conversely, tolerant alleles seem to be outcompeted off the mine, indicating that there they are less fit. In other words there appears in this case to be a fitness cost of tolerance.

Figure 13.8 also contrasts the copper tolerance of individuals grown from seeds (white bars) with that of adults (black bars). Comparison of the two shows how selection works at different places on the transect. Thus adults on the mine are more tolerant than seeds, indicating selection for tolerance on the mine. Conversely, downwind of the mine adults are less tolerant than seeds, indicating selection against tolerance off the mine.

It is generally believed that there is a fitness cost of tolerance, i.e. that tolerant genes are disadvantageous under normal conditions. If this were not the case, then tolerance genes would occur widely, at least in some species, given their competitive edge in polluted environments. However there is disagreement as to the severity of the cost of tolerance. Some authors consider they may be great (see Baker, 1987). Others suggest they may be small, on the basis of studies which carefully distinguish metal tolerance from other evolved attributes of mine populations (Macnair, 1987). For example, tolerant strains may be those which grow more successfully in the nutrient-poor conditions often found in mine waste (Ernst *et al.*, 1992).

Ernst (1976) suggested that the fitness costs of tolerance are a result of the energy needed by the mechanisms of tolerance. Energy spent on detoxication is not available for growth, and Ernst suggested this is why many tolerant plants grow more slowly and produce less biomass than non-tolerant conspecifics. This is the situation illustrated in Figure 13.2A.

Why do some species evolve metal tolerance and others not? Presumably similar selection pressures apply to all. Selection alone, however, is not enough, there must also be present some suitable 'tolerance genes' on which selection can act. Table 13.3 shows that normal populations of species able to evolve metal tolerance contain tolerance genes at low frequencies. Theory shows that whether or not tolerance genes exist in normal populations depends on mutation rate, i.e. the rate of creation of tolerance genes by mutation, on the fitness cost of tolerance, and on the population size. Tolerance is more likely to evolve if the mutation rate is high, the fitness cost of tolerance low, and the species is common.

13.6.3   *Evolution of Industrial Melanism*

The blackening of industrialized parts of the countryside was a result of the Industrial Revolution which began in Britain in the 18th Century. One result was that an area west of Birmingham became known as the Black Country. Blacker ('melanistic') forms of animals are better camouflaged in such environ-

Table 13.3 The percentage of copper-tolerant individuals found in normal populations of various grass species that are commonly found near mines in Britain, in relation to whether copper-tolerant populations of these species have been found on copper mines[a]

Species	Occurrence of tolerant individuals in normal populations (%)	Presence (+) or absence (−) of tolerant populations on copper mines
Holcus lanatus	0.16	+
Agrostis capillaris	0.13	+
Festuca ovina	0.07	−
Dactylis glomerata	0.05	+
Deschampsia flexuosa	0.03	+
Anthoxanthum odoratum	0.02	−
Festuca rubra	0.01	+
Lolium perenne	0.005	−
Poa pratensis	0.0	−
Poa trivialis	0.0	−
Phleum pratense	0.0	−
Cynosurus cristatus	0.0	−
Alopecurus pratensis	0.0	−
Bromus mollis	0.0	−
Arrhenatherum elatius	0.0	−

[a] Data of C. Ingram reported in Macnair (1987).

ments, and may thus be better protected against predators. This may give melanistic forms a survival advantage. Melanism is found in many species of arthropod, but most examples are moths. The account here is based on Brakefield (1987) and Moriarty (1988).

The incidence of melanism has risen steadily since about 1850. Over 100 of the 780 species of larger moths in Britain now commonly include melanic forms. Similar changes have occurred in Europe and North America, usually in industrial areas. In many cases the melanic forms have spread very rapidly and have become predominant.

The first and best studied example is the peppered moth, *Biston betularia*, (Kettlewell, 1973). J. W. Tutt suggested in 1896 that the non-melanic form is well concealed ('cryptic') when at rest on the pale bark of trees in rural areas, where it resembles thalli of foliose lichens. By contrast it was conspicuous on the completely black surfaces of trees and walls in industrial Britain. The blackening was mostly due to soot, but lichens were also killed, mainly by sulphur dioxide. Tutt suggested that in this environment the non-melanic form was conspicuous to avian predators, conversely melanic forms were cryptic.

The first melanic specimen was caught in Manchester in 1848, and by 1895 98% of the moths in that area were melanic. This corresponds to a 50% increase in the chances of survival of the melanic forms, according to calculations by J. B. S. Haldane (see Brakefield, 1987). Increases in melanic forms

257

were recorded during this period in a number of industrialized zones. The phenomenon has been of major interest to evolutionary biologists since the 1920s.

Kettlewell organized surveys between 1952 and 1970 of the geographical distribution of melanism in *B. betularia* in Britain (Figure 13.9). Three forms of the moth were recognized. The non-melanistic form, *typica*, shows a range of coloration, from heavily speckled individuals to others that are almost white with fine, granular black markings. The common melanic form, *carbonaria*, produced by a dominant allele at a single locus, is usually all black, but sometimes has some light-coloured spots or patches. A third, intermediate, form, *insularia*, is also recognised. Figure 13.9 shows that in general the melanic forms occurred in the industrialized regions of Britain, and in the areas downwind (the predominant winds in Britain are from the southwest). Kettlewell

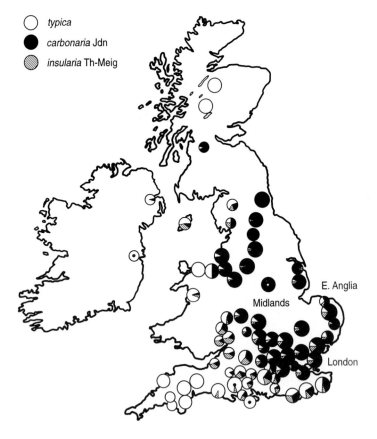

Figure 13.9 The relative frequencies of the normal and two melanic forms of the peppered moth, *Biston betularia*, in Britain. The results are based on more than 30 000 records collected from 1952 to 1970 at 83 sites. From Kettlewell (1973).

suggested that the downwind occurrence of the melanistic forms might be because soot had blackened the downwind areas, but it could also be the result of passive wind dispersal of larvae. When eggs hatch, the very small larvae suspend themselves on silk threads and are dispersed by air currents. It is possible that the high incidence of melanism in East Anglia is the result of long-distance dispersal of larvae from industrial zones in London and the English Midlands.

Interestingly, there has been a further environmental change in recent years consequent on clean air legislation and the creation of smokeless zones in the late 1950s. This led to a rapid fall in emissions of smoke and sulphur dioxide. Tree surfaces became lighter, and the relative frequency of melanics fell (Figure 13.10). This phenomenon has been described as 'evolution in reverse'! The time lag between the reduction in pollution and the evolutionary response is probably a result of slow change in the composition of the lichens covering the tree branches on which the moths rest.

Although it is generally accepted that melanism in *Biston betularia* is a result of atmospheric pollution, the detailed working of the selective mechanisms is still not fully understood. The fitness advantages, of the melanic form in industrial areas and of the non-melanic forms in rural areas, were demonstrated by Kettlewell in mark-release-recapture experiments. Marked individuals of melanic and normal forms were released in rural and industrial areas (Table 13.4). More melanic individuals were recaptured in the industrial area, and more non-melanic in the rural areas.

To go further and to estimate the predation rate by birds it is necessary to have a detailed knowledge of the time budgets of adult moths of different ages and both sexes during the reproductive period. Peppered moths emerge shortly before dusk. After a dispersal flight females rapidly attract mates, and

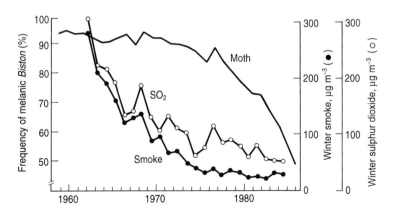

Figure 13.10 The decline in the frequency of the melanic form of the peppered moth near Manchester after clean air legislation reduced emissions of smoke and sulphur dioxide. From Brakefield (1987).

Table 13.4 The relative recoveries of marked individuals of two forms of the peppered moth, *Biston betularia* (*typica* and *carbonaria*) from two sites, one rural and one industrial[a]

Site	Form	No. released	No. recaptured	Recapture (%)
Rural Dorset	*typica*	496	62	12.5
	carbonaria	473	30	6.3
Industrial Birmingham	*typica*	201	32	15.9
	carbonaria	601	205	34.1

[a] From Moriarty (1988) presenting data from Kettlewell (1955, 1956).

pair shortly after dusk. Thereafter they do not fly and only walk short distances. They remain *in copula* for nearly 24 h. Many moths rest during the day underneath or on the side of small branches in the tree canopy. Studies of visual predation on the melanic and non-melanic forms have however generally been carried out on tree trunks. Although the findings are qualitatively in keeping with the camouflage-protection hypothesis, the experiments need repeating on more realistic substrates.

13.6.4 Evolutionary Response of Dog Whelks, Nucella lapillus, to TBT Contamination

The physiological and ecological effects of marine antifouling paints containing TBT were described in Chapter 11, section 11.4.3. To recapitulate, female dog whelks suffer 'imposex', a condition involving the growth of male sex organs. This can prevent successful reproduction, and because migration is very limited in this species, sterilization by imposex has resulted in its extermination in many areas of its European distribution, particularly shores adjacent to commercial harbours and pleasure boating centres where water TBT levels have exceeded 2 ng Sn l^{-1}. One such area is the north coast of Kent, where dog whelks are known to have been abundant in pre-TBT times. Recently, however, Gibbs (1993) discovered that a population in this area at Dumpton Gap had evolved modified genitalia that allowed it to persist. The account here is based on Gibbs' paper.

The Dumpton Gap population is characterized by the absence of a penis or an undersized penis in about 10% of the males (absence of a penis in males is unknown elsewhere). The vas deferens and prostate are incompletely developed in affected males. Gibbs has labelled these abnormalities the 'Dumpton syndrome'. Laboratory-bred animals display the same characteristics, suggesting that the character is genetically determined. It cannot be due to some aberrant feature of the Dumpton environment such as predators or parasites, since neither predators nor trematode parasites were present in the laboratory. About 75% of the Dumpton Gap females showed little or no imposex, whereas usually all females are expected to show imposex at the TBT levels experienced at Dumpton Gap.

The evidence therefore suggests that the 'Dumpton syndrome' is the result of a genetic mutation which reduces imposex in females, allowing them to breed successfully, even though it prevents breeding in some 10% of males. Overall, the gene must confer a marked fecundity advantage in environments contaminated with TBT. In TBT-free environments, however, the gene would be disadvantageous since a significant proportion of males are infertile (compare Table 13.1). It seems, therefore, that the 'Dumpton syndrome' carries a fitness cost, and so would be selected against in TBT-free environments.

13.7 Conclusions

Several points emerge from these case studies. They are good studies of the evolutionary process because they report evolutionary responses to known environmental changes. In each case the change is chronic man-made pollution which started at most a few centuries ago. For evolutionary responses to occur, genetic variation has to be present on which selection can act. Evolutionary responses occur when the change to the environment alters the relative fitnesses of different alleles. In each case there was strong selection for resistant alleles in polluted environments, and evolutionary responses to pollution occurred within tens of generations. Where the genetic mechanisms have been investigated, they consist for the most part of one or occasionally two major genes. Although it is extraordinarily hard to study, it seems likely that resistance generally entails a fitness cost. In other words resistant alleles are favoured at polluted sites, but selected against at unpolluted sites. Lastly it should be emphasized that although our outline understanding seems secure, in each case there is still some uncertainty as to the detailed working of the evolutionary process.

Further Reading

The account of the evolutionary process given here is based on SIBLY and ANTONO-VICS (1992). This approach was selected because it is accessible and gives insight into how selection acts on life-history characters such as juvenile growth rate, fecundity and survival. Those wanting a population genetic text should consult HARTL and CLARK (1989); for a general evolutionary text see RIDLEY (1993). BISHOP and COOK (1981) remains a useful review of the genetic consequences of man-made environmental changes.

MALLET (1989) gives a brief but accessible introduction to the evolution of pesticide resistance.

ROUSH and DALY (1990) provide a heavyweight authoritative review, invaluable to any one starting work on any aspect of resistance.

MACNAIR (1987) gives a short but useful introduction to the evolution of heavy-metal tolerance in plants; SHAW (1989) contains a collection of reviews of this field.

14

Changes in Communities and Ecosystems

14.1 Introduction

In the two preceding chapters, effects upon individual species were given particular attention. This is the approach usually followed when dealing with larger species. By contrast, microbiologists are particularly concerned with effects on communities and ecosystems, a subject area sometimes termed 'synecology'. The measurement of effects of pollutants upon ecosystems has certain strengths and limitations. It has the advantage of being a holistic approach, which can take into account the overall functional state of an ecosystem. Also, the analytical methods that are employed can be relatively simple and inexpensive e.g. when measuring CO_2 production or nitrification in soils. On the other hand, effects upon individuals and, thus, on the composition of communities may pass undetected. For example the addition to soil of a herbicide can lead to the upsurge of certain species or strains of microorganisms which can use the chemical as an energy source; but this may not be reflected in any detectable change in the rate of CO_2 production by soil.

An ecosystem approach has been used with some success to study the effects of pollutants upon soils. The features of soil communities that make them amenable to scientific study are that (1) the boundaries of communities can be easily and clearly defined, and (2) similar communities exist in vast numbers in the field. These features make it possible to investigate the factors determining their field distribution, and to carry out field experiments.

These useful attributes of soil communities are not shared with most larger-scale communities and ecosystems. Exceptions are the aquatic communities inhabiting lakes and rivers. Lakes exist in enormous numbers in Canada, Siberia and Scandinavia. The major types of pollution are acidification and heavy-metal pollution, and their effects can be readily observed in terms of loss of species and diversity (see below, section 14.3). Field experiments are in principle a possibility, although because of their scale they are expensive.

Other types of terrestrial and aquatic communities are not so amenable to study. This may be because either their constituent species are highly mobile,

263

so that their boundaries are hard to define, or because no two pollution events are the same. For instance severe oil pollution rarely affects the same type of community twice, and the boundaries of the affected marine communities are ill-defined. Similar consideration applies to air pollution, or radioactive pollution as a result of nuclear warfare or accidents at nuclear power stations.

In the extreme case of an ecosystem analysis, the composition of the atmosphere can be taken as an index of the state of health of the entire planet. The recent increase in CO_2 levels in the atmosphere gives evidence of pollution on the global scale. This presents the threat of global warming because of the so-called 'glasshouse' effect. Likewise, the disappearance of the ozone layer above Antarctica has been attributed to the appearance of CFC gases in the stratosphere. Reduction of the ozone layer brings an increase in ultraviolet radiation reaching the earth's surface, and a consequent threat of cellular damage to living organisms (e.g. skin cancer in humans). In both of these examples there is a suggestion that ecological damage on the global scale may result from changes in natural processes caused by pollutants – viz. the carbon cycle and the oxygen/ozone cycle. This view of pollution in terms of 'global' ecology, is discussed by Lovelock in his elaboration of the 'Gaia' theory (see section 14.4).

14.2 Soil Processes

Soil communities are complex associations between a variety of micro- and macroorganisms, minerals and non-living organic materials (refer to Chapter 5). The carbon cycle and the nitrogen cycle operate in soils and both provide indices of the function of soil communities. Stages in the cycles, e.g. CO_2 production (carbon cycle) and nitrification (nitrogen cycle), can be measured in the presence and absence of pollutants.

The basic carbon cycle, shown in Figure 14.1, is an example of a nutrient cycle. Organisms which obtain their carbon from organic compounds are termed 'heterotrophs'. Animals and many microorganisms fall into this category. Organisms which obtain their carbon from carbon dioxide are termed 'autotrophs'. Autotrophs include green plants, green algae and certain bac-

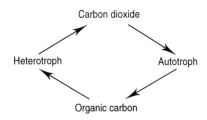

Figure 14.1   The carbon cycle.

teria, and play a vital role in fixing atmospheric carbon dioxide into organic compounds. The heterotrophs then complete the cycle by converting organic compounds to carbon dioxide. Thus, the operation of the cycle may be affected by the action of pollutants upon autotrophs, heterotrophs or both.

A simplified version of the nitrogen cycle in soil is shown in Figure 14.2. Atmospheric nitrogen can be 'fixed' by certain bacteria, including both free-living species (e.g. *Azotobacter* spp) and symbiotic bacteria (e.g. *Rhizobium* species, which are found in the root nodules of leguminous plants). Fixation involves conversion of molecular nitrogen to ammonia, which then forms ammonium ions (NH_4^+) in soil water. NH_4^+ can be oxidized sequentially to nitrite ions (NO_2^-), and nitrate ions (NO_3^-) by soil bacteria (nitrification). Plants and bacteria can take up NH_4^+ and/or NO_3^- and use them for the biosynthesis of organic nitrogen compounds (e.g. amino acids and purines). Heterotrophs can then utilize organic nitrogen compounds as sources of nitrogen, and release NH_4^+ and NO_3^- from them. Finally, some microorganisms can convert NO_3^- to nitric oxide (NO), nitrous oxide (N_2O) and nitrogen, a process termed denitrification.

Microorganisms have an important role in the operation of the carbon and nitrogen cycles. Under normal environmental conditions these cycles are subject to the influence of environmental factors such as temperature, pH and soil water content. The degree to which these environmental factors vary in time and space is dependent upon geographical location and the inherent properties of the soil (e.g. clay content, organic matter and $CaCO_3$ content). In evaluating the effects that pollutants may have on soil processes, it is important to see them in relation to the fluctuations that occur in unpolluted sites. Effects of pollutants can be regarded as serious and significant if they go beyond the normal variations in operation of the cycles.

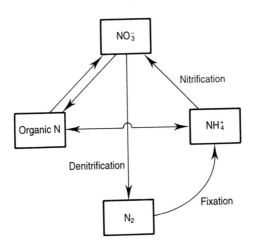

Figure 14.2 The nitrogen cycle.

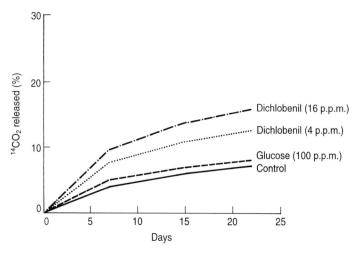

Figure 14.3 Effect of a herbicide (dichlobenil) on the rate of CO_2 production in soil. ^{14}C-glucose was added to soil as a carbon source for microorganisms. The addition of the herbicide dichlobenil caused an increase in the rate of release of $^{14}CO_2$ (derived from ^{14}C-glucose) over a 22-day period. The rate of $^{14}CO_2$ release from soil is expressed as the percentage added ^{14}C which appears in this form. From Somerville and Greaves (1987).

A widely used method of estimating hazards of chemicals to micro-organisms involves determination of the *rate of carbon dioxide formation* in soil samples. Usually a source of carbon (e.g. plant or horn meal) is added to soil to stimulate CO_2 formation, and a comparison is made between soils with and without the test chemical (Figure 14.3).

The rate of production of nitrite and nitrate from soil ammonium (*nitrification*) can also be readily measured (Table 14.1). This is regarded as a valuable method of testing the effects of pollutants because of the importance of nitrification in relation to soil fertility. A further example is measurement of

Table 14.1 Effects of pollutants upon nitrification in soil[a,b]

Treatment	mgNO$_3^+$ N/day	% of control
Control	0.53	100
5 × normal application rate of a pesticide	0.54	102
1 p.p.m. Nitrapyrin	0.14	26

[a] Horn meal was added to the soil as a nitrogen source. The rate of production of nitrate was unaffected by a high application of a pesticide. It was, however, strongly inhibited by nitrapyrin, a bactericide, used as a positive control.
[b] From Somerville and Greaves (1987).

the rate of nitrogen fixation by root nodule bacteria in a leguminous crop, which may be affected by soil pollutants.

Another factor to be considered is the *duration* of any effect produced by a chemical. Pollutants can change the composition of communities of soil microorganisms (Chapter 5). When a chemical has an immediate effect on soil processes due to chemical toxicity, there may be a subsequent population growth in the species/strains of microorganisms which can metabolize it and use it as a nutrient source. Thus, the effects of an organic chemical will be relatively short lived because the increase in numbers of these microorganisms will lead to a more rapid breakdown of the compound in soil. It has been suggested that effects of chemicals lasting up to 30 days be regarded as 'normal' (1), up to 60 days as 'tolerable' (2), and beyond 60 days as critical (3). In reviewing effects of pesticide on soil microflora, some 90% of all cases fell into the first category.

A general problem when performing soil tests to evaluate the effects of chemicals is choosing the type of soil to use, and the appropriate operating conditions. Attempts have been made to define standard soils – but the trouble here is that these are only representative of particular geographic and climatic areas. Also, a standard soil may not represent a 'worst case' scenario – it may not be the soil most likely to show the effect of a particular chemical.

Soil tests of the type described are of particular interest in regard to the testing of pesticides. Herbicides, fungicides, and insecticides are chemicals with high biological activity which may be expected to have effects on soil organisms and consequently upon soil fertility.

14.3   Acidification of Lakes and Rivers

The pH of surface waters is determined by both abiotic and biotic factors. Such factors as the composition of the base rock, and the precipitation of acids in rain or snow, have a strong influence upon pH. Also important are the release of acids when organic residues are decomposed by microorganisms, and the influence of neighbouring forest land. pH, in turn, influences the composition of aquatic communities. In particular, reduction of pH below 6 can be harmful to many species. Thus, the pH of surface waters is affected both directly *and indirectly* by pollutants (pollutants may affect pH of surface waters indirectly through action upon microorganisms or plants). Thus, the pH of surface waters is dependent upon the operation of natural processes, as well as pollution, and provides an indication of the diversity and 'health' of aquatic ecosystems.

The deposition of acid rain has resulted in the acidification of weakly-buffered surface waters in many areas, including Scandinavia, Eastern Canada and the Northeastern United States. For example, the pH of 21 water bodies in Central Norway decreased from an average of 7.5 to 5.4–6.3 between 1941 and the early 1970s; the pH of 14 surface waters in Southwestern Sweden decreased from 6.5–6.6 prior to 1950 to 5.4–5.6 in 1971; the average pH of seven rivers in Nova Scotia decreased from 5.7 in 1954–1955 to 4.9 in 1973. Although doubts have been expressed about the accuracy of older pH data, there is a broad scientific consensus that there has been recent acidification of many surface waters. The material in this section is based in part on Freedman (1995).

The effects of acidification on aquatic communities are sometimes dramatic, as in the loss of fish populations from a number of highly acidified waters. For instance, commercially-valuable salmonids have been lost from many surface waters in Scandinavia (example in Figure 14.4). A survey of more than 2000 lakes in Southern Norway showed that one-third have lost their fish populations since the 1940s. The fishless lakes generally had a pH < 5.0. The local salmonids vary in their sensitivity to pH; requirements range from pH > 4.5 for non-migratory brown trout (*Salmo trutta*) to 5.0–5.5 for Atlantic salmon (*Salmo salar*). Juveniles are generally more sensitive than adults, but adult deaths are sometimes recorded in Spring when water pH is lowest. Similar results involving more species have been recorded in many studies in Canada and the Northeastern United States.

Field experiments are possible with lakes, just as they are with soil communities, but the costs are greater. The best example is the experimental acidification of lake 223, a 27-ha oligotrophic lake in the 'Experimental Lakes Area' of Northwestern Ontario. The lake was extensively studied in 1974 and 1975 to provide baseline data, and then the lake was progressively acidified so that pH fell from 6.5–6.8 to 5.0–5.1 by 1981, after which pH was maintained in the range 5.0–5.1 until 1983. This was achieved by adding sulphuric acid, 27 400 litres having been added by 1983. Experimental acidification has subsequently been carried out at two other lakes, one in the Experimental Lakes Area, the other in Northern Wisconsin. These lakes have different physical settings and originally had different assemblages of plants and animals, including fish.

Despite their initial differences, the responses of these three lakes to acidification were in many respects remarkably similar, particularly with regard to biogeochemical processes and effects on the lower trophic levels (Schindler *et al.*, 1991). Although acidification disrupted nitrogen cycling in all three lakes, each generated some buffering capacity internally.

The phytoplankton communities of all three lakes had originally been dominated by chrysophyceans and cryptophyceans. Acidification changed the dominant species and decreased diversity. Phytoplankton production and standing crop were somewhat increased, probably because light penetration increased. The littoral zones of the lakes, however, became dominated by a few species of filamentous green algae. These formed mats or clouds of algae which changed the entire character of the littoral zones when the pH fell below 5.6.

Acidification also changed the zooplankton communities, with cladocerans becoming increasingly dominant. As acidity increased, *Daphnia catawba* took over from other *Daphnia* species; the identity of the dominant rotifer species changed, and several sensitive zooplankton species disappeared.

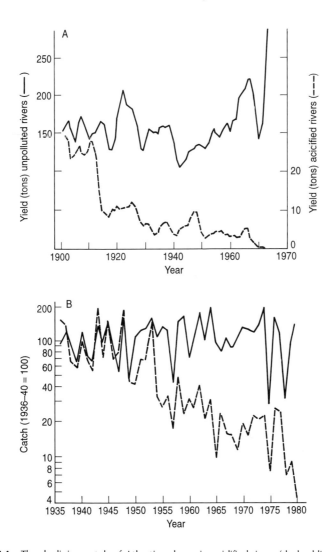

Figure 14.4 The declining catch of Atlantic salmon in acidified rivers (dashed lines) compared with less polluted rivers (solid lines). (A) In southernmost Norway: seven acidified rivers compared with the rest of the country (modified from Leivestad *et al.* (1976). (B) In Nova Scotia: 12 rivers with pH > 5.0 compared with 10 with pH ≤ 5.0 in 1980 (from Watt *et al.*, 1983).

Acidification also caused the loss of several species of large benthic crustaceans. The responses of fish species, however, were varied and appeared to depend on the sensitivity of key organisms lower down the food chains.

When acidification was reversed in lake 223 between 1984 and 1989, there was rapid partial recovery of the lacustrine communities, though the

269

assemblages of phytoplankton and chironomids retained an acidophilic character.

Overall acidification consistently reduced diversity, through loss of species and through the increased dominance of a small number of acidophilic taxa. These responses are similar to those found in atmospherically-acidified lakes.

14.4 Global Processes

The atmosphere enshrouds the whole planet and provides a link between oceans and continents. The air masses that constitute the lower atmosphere are in a constant state of circulation, individual gaseous components moving by diffusion processes through the air masses (Chapter 3). The composition of the atmosphere is influenced by living processes at the earth's surface. Levels of carbon dioxide, oxygen and nitrogen are dependent upon living processes. Thus, the effects of pollutants on ecosystems can, at least in theory, influence the operation of these processes globally and thus alter composition of the atmosphere. Some scientists conceive of the entire earth as being a single living entity (as for instance, in the Gaia theory of Lovelock).

If looking at things in this way, pollutant effects on living organisms may be direct or indirect. Inhibitors of photosynthesis are examples of pollutants which act directly on plants, slowing down the use of CO_2 and the release of O_2. The action of CFC gases is indirect. Due to a reduction of the ozone layer, more ultraviolet radiation reaches the earth, causing damage to animals and plants.

The level of CO_2 in the atmosphere has been the subject of much debate. Over the last 40 years this has increased by about 10%. It is thought that most of this increase is due to the combustion of fossil fuels (coal, oil etc.) It has also been argued that a substantial reduction in the area of forest land has also contributed to this – since trees are responsible for 'locking up' considerable quantities of the gas. The concern over the rise of atmospheric CO_2 is that it is having a 'glasshouse effect' – retaining more heat at the earth's surface and thereby causing global warming.

The issues raised here are long-term ones which are unlikely to go away. The effects of pollutants on the global scale can influence the state of ecosystems over the entire planet, and are likely to be of increasing concern into the next century.

Further Reading

FREEDMAN, B. (1995) Excellent coverage of the effects of pollution on ecosystems, and, in particular, useful for the effects of acidification on aquatic ecosystems.
SCHINDLER *et al.* (1991) A review of the experimental acidification of lakes.
LOVELOCK, J. (1989) *The Ages of Gaia*. An exposition of the Gaia hypothesis.
WOOD, M. (1995) *Environmental Soil Biology*. Gives an account of the soil processes that may be affected by pollutants.

15

Biomarkers in Population Studies

It could be argued that the most crucial task for ecotoxicologists is to ensure that the structure and the function of ecosystems are preserved. It is also the most difficult. The linkages between biochemical, physiological, individual, population and community responses to pollutants are shown diagrammatically in the Introduction. The dilemma is that as the importance of a change increases so does the difficulty of measuring it and relating it to a specific cause. In this chapter a number of specific examples will be considered to illustrate the approaches that have been used to tackle this important problem of ecotoxicology, involving the use of biomarkers. These examples are as follows:

(1) DDE-induced eggshell thinning in raptorial and fish-eating birds;
(2) reproductive failure of fish-eating birds on the Great Lakes of North America;
(3) reproductive failure of molluscs caused by tributyl tin;
(4) the forest spray programmes of Eastern Canada.

The first three are investigations of unexpected adverse side-effects of the use of pesticides and/or industrial chemicals, and the fourth is an investigation of a large-scale operational use of pesticides.

15.1 DDE-induced Eggshell Thinning in Raptorial and Fish-eating Birds

In the early 1950s, Derek Ratcliffe of the British Nature Conservancy found what he considered to be an abnormal number of broken eggs in the eyries of the peregrine in several regions of the British Isles. Later in that decade, representations were made to the British Home Office by racing pigeon enthusiasts concerning the losses caused by peregrines preying on their birds. These persons were lobbying for a change in the protected status of the peregrine because they claimed that the peregrine population had greatly increased. As a result of this pressure, the Nature Conservancy was asked to undertake a nation-wide survey. This revealed that the peregrine population had declined greatly and that less than one-fifth of the birds successfully raised young in 1961 and 1962 (see Figure 12.15).

This paper had an immediate impact in America. A team surveyed peregrines in Eastern North America, travelling 22 500 km and visiting 133 known eyries in 1964 and found every single eyrie deserted. This finding was the spur for a meeting to examine the population changes of the peregrine and other birds-of-prey. The meeting was held at Madison, Wisconsin, in 1965 (Hickey, 1969) and was attended by persons from many parts of the world interested in the peregrine. The data presented at the meeting showed that the species was extirpated in Eastern North America and had decreased markedly in many countries in Europe. In Scandinavia, the population was reduced to only 5% of its pre-war population by the early 1970s. In Germany the number of breeding peregrines had decreased from 400 pairs to 40 pairs by 1975.

Based on the finding that egg breakage had become more common, it was likely that the thickness of eggshells had become thinner. Ratcliffe examined the temporal variation in peregrine and sparrowhawk eggshells collected in the UK since 1900. His findings, that a pronounced decrease in eggshell thinning occurred in both species in the mid-1940s, were published in *Nature* in 1967. A replotting of Ratcliffe's data for the peregrine, over the critical period of change, is shown in Figure 12.15. The first sign of a change was in 1946, although the mean was not statistically significant. In 1947 the change was clear and statistically significant, and in the decade that followed eggshells as thick as the pre-war norm were a rarity.

The most dramatic case of eggshell thinning was found in the brown pelican (*Pelecanus occidentalis*) off the coast of California in 1969. The colony on Anacapa Island showed almost complete reproductive failure, with only four chicks raised from some 750 nests. Most nests were abandoned and remains of crushed eggshells were found throughout the colony (see Figure 15.1A). On average, the thickness of these broken and crushed eggshells was only half of the normal value. The reproductive failure of the colony continued for the next few years. The productivity of double-crested cormorant (*Phalacrocorax auritus*) colonies was also close to zero and, again, the main cause was the breakage of the eggs. The most detailed studies on eggshell breakage were made by the Canadian Wildlife Service on colonies of cormorants on Lake Huron in 1972 and 1973 (Weseloh *et al.*, 1983). These workers visited all the known colonies of cormorants in Lake Huron. They found that 79% of the eggs were 'lost' within 8 days of laying and that by the end of the normal incubation period only 5% of the eggs remained in the nests. In about half the cases of 'lost' eggs, eggshell fragments were found in or around the nest. The eggshells averaged 24% thinner than pre-war values; while this is not as severe as that found in California it was enough to cause almost com-

A

B

Figure 15.1   (A) Crushed eggs in the nest of a brown pelican, Anacapa Island, California, 1970. (B) Double-crested cormorant with deformed bill from a colony on Lake Michigan, USA.

273

plete reproductive failure. The subsequent recovery of this population is detailed in section 15.2.

The importance of the North American findings was that there clearly was a linkage between reproductive failure and eggshell thinning. Analytical work revealed that DDT (and its metabolites) and PCBs were the only contaminates present in appreciable amounts. In this, the situation is quite different to that in the UK where dieldrin (see Chapter 12) was found to be a major factor in population declines of peregrine and sparrowhawk.

It was thought that a pesticide was responsible for eggshell thinning on the grounds that the effect occurred at the time that synthetic pesticides were introduced on a large scale and the declines in population were greatest in areas where pesticides were most heavily used. The analysis of residues of organochlorine pesticides was becoming more widespread by this time and by the beginning of the 1970s it was possible to show a correlation between the levels of DDE in the egg contents and the thickness of the shell (Figure 15.2). This particular dose–response curve shows that, initially, a low concentration of DDE causes a large response and that the response tends to 'flatten out' with increasing dose.

These findings established a correlation between the concentration of DDE and the degree of eggshell thinning. Experimental studies with the American kestrel established a cause-and-effect relationship. Further, it was demonstrated that the relationship between the degree of eggshell thinning and DDE residue levels in the egg based on laboratory studies was the same as that found in the field. These studies also showed that PCBs did not cause eggshell thinning.

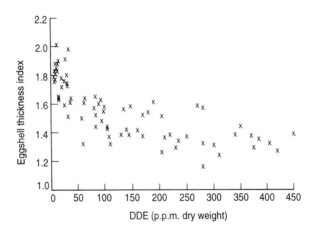

Figure 15.2 Relationship of eggshell thickness index to DDE residue levels in peregrine eggs collected from Alaska and Northern Canada. From Peakall (1993), reprinted with permission from *Environmental Reviews*.

There was a major legal battle over DDT in 1971. It was the first pesticide over which environmental regulations were made and the producers feared, with some justification, that if they lost this battle they would lose others. The hearing that led to the ban of DDT in the United States was the most extended and bitterly fought ever conducted on an environmental contaminant. It ran from August 1971 to March 1972 and a total of 125 expert witnesses produced over 9000 pages of testimony. There were four batteries of lawyers. On one side there were those representing the chemical industry (27 companies acting together as group petitioners), and the lawyers of the US Department of Agriculture; on the other side those representing the US Environmental Protection Agency and the Environmental Defence Fund.

Observations of eggshell thinning linked to DDT were important evidence in this hearing. Despite the verdict going against DDT (and other similar ones in other countries), there was criticism of the finding. Because eggshell thinning had occurred so rapidly after the introduction of the insecticide, before it was widely used, it was argued that DDT could not have caused this effect. It was indeed true that all the work was done some 10–20 years after thinning had first been shown to occur; the first measurements were made on the levels of DDE in peregrine eggs in 1962 whereas eggshell thinning started in 1946. However, so stable is DDE and so sensitive are analytical techniques that it proved possible to extract enough DDE from the dried membranes of eggs collected over the critical period (1946–47) to demonstrate that enough DDE was present to have caused eggshell thinning.

It was soon found that the phenomenon of eggshell thinning in peregrines was global and that there was a close correlation between the degree of eggshell thinning and the health of the peregrine population. Studies from Australia, Europe and North America showed that when eggshell thinning exceeded 17–18% population declines followed. The relationship between eggshell thinning and the status of populations is shown in Figure 15.3. Four extirpated populations had eggshell thinning of 18–25%, declining populations, over 17%, whereas stable populations had less than 17% thinning.

As would be expected, there was considerable interest in the mechanism whereby DDE caused eggshell thinning. A wide variety of mechanisms have been proposed. However, the finding that eggshell thinning of up to 50% occurred in the brown pelican indicates that the site of impairment is in the shell gland, as reduction of 50% of calcium levels throughout the organism would cause profound physiological changes. This hypothesis is supported by the finding that the circulating level of calcium in the blood is normal in birds laying eggs with thin shells. There is now general agreement that it is transport of calcium across the eggshell gland mucosa that is affected by DDE. Decreased activity of Ca-ATPase and effects on prostaglandins have been proposed as the key mechanisms. There is no reason to suppose that these mechanisms are mutually exclusive. However, no comprehensive answer to the

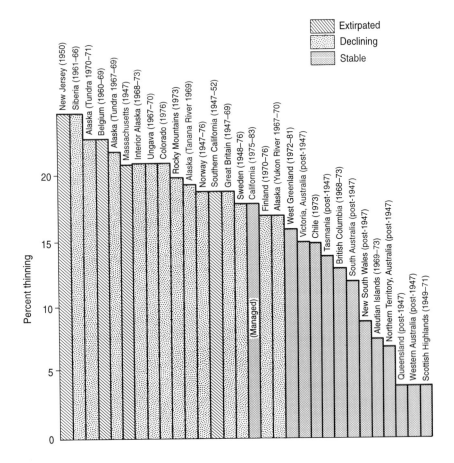

Figure 15.3 Relationship between degree of eggshell thinning and the status of various populations of peregrines. From Peakall (1993), reprinted with permission from *Environmental Reviews*.

mechanistic basis of species variation of DDE-induced eggshell thinning has been put forward.

The evidence that DDT and dieldrin was causing widespread decreases in several populations of the peregrine and many other birds of prey was part of the evidence that led to bans and restrictions. The use of DDT and dieldrin was banned in the Scandinavian countries over the period 1969–72, and in Holland dieldrin was restricted in 1968, and DDT was banned in 1973. In Great Britain there were voluntary restrictions on both DDT and dieldrin starting in 1962, although some small-scale uses were not banned until 1986. The approach in Great Britain was the opposite to that in the United States, where the ban on DDT imposed in 1972 followed a long court hearing.

In the Eastern United States and Southern Canada, the peregrine had disappeared completely by the early 1960s. In these areas, re-introduction was the only feasible means of restoring the population. Large-scale breeding programmes, with the intention of raising birds for re-introduction, were started by the Canadian Wildlife Service at Wainwright, Alberta, in 1972 and by the Peregrine Fund at Cornell University, New York, at about the same time. The first few captive-raised young were released in 1975 and the numbers released soon increased. The first breeding attempt of released birds in the wild was in 1979 and the progress has been rapid since then. We have now reached the point where releases are needed only in a few specific areas.

In other parts of the world, some breeding stock remained and re-introduction programmes were not essential. Recovery has occurred in Alaska, Western Canada, the continental United States and Mexico, starting in the late 1970s. In the British Isles, where the most detailed data are available, recovery was described as 'virtually complete in overall numbers, though not in precise distribution'. In other parts of Europe, the populations have increased, but are still low compared to pre-war numbers, notably in the Czech Republic, Slovakia, East Germany and Poland, which formerly held large populations. These, and other, recoveries of the peregrine are detailed in Cade *et al.* (1988).

An interesting aspect of the phenomenon of eggshell thinning is the wide difference in the sensitivity of different avian species. Most sensitive are the raptors and the fish-eating birds; this is unfortunate as it is these species that are exposed to the highest dose due to the bio-accumulation of DDE up the food chain. The raptors, in addition to the peregrine, that have shown marked effects include the osprey (*Pandion haliaetus*) and bald eagle (*Haliaetus leucocephalus*) in North America and the sparrowhawk in Europe. In contrast, the species most commonly used in *in vivo* toxicity tests (quail, pheasant, chicken) are almost completely insensitive; and others (such as the duck) are only moderately sensitive. Thus, it is unlikely that even the studies conducted now prior to registration of a new pesticide would have detected this particular adverse effect of DDT. However, other negative aspects of DDT, such as its strong tendency to biomagnify up the food chain and its toxicity to fish, should prevent registration.

No story is quite as tidy as one would like. Although in many parts of the world it seems clear that DDT-induced eggshell thinning was the main cause of population declines of raptorial birds, in some areas other pesticides were certainly also involved. In the United Kingdom direct mortality caused by dieldrin, another organochlorine that is magnified up the food chain, is considered to have been the most important factor (see Chapter 12), whereas in North America the levels of dieldrin recorded were well below the critical value.

The eggshell thinning story is one of the most comprehensive environmental investigations in the short history of environmental toxicology. An important feature of eggshell thinning and the subsequent collapse of the populations of the peregrine and other birds-of-prey lies in the fact that the effects occurred over a wide area, much of it remote from the place of application of the pesticide.

15.2 Reproductive Failure of Fish-eating Birds on the Great Lakes of North America

The North American Great Lakes are, collectively, the largest body of freshwater in the world. The very vastness of the Great Lakes was strangely one of the causes of their pollution. They are so large that 'dilution seemed the solution to pollution'. The waterway of the lakes was a major factor in the opening up and development of this part of the New World. Hardly surprisingly, towns and industry sprang up on their shores. The harnessing of the Niagara Falls to generate electricity led to major industrial development along the Niagara River. Now some 36 million people live within the Great Lakes basin and over 13 000 manufacturing and industrial plants are located there. The result was that, by the 1960s, serious wildlife problems were occurring on these inland seas.

Investigations of the problems with wildlife of the Great Lakes had two separate and very distinct beginnings. One was proactive and the other reactive. The proactive approach was lead by Joe Hickey, Professor of Wildlife Ecology at the University of Wisconsin. Concern over DDT was rising at this time and the study was started 'to determine what pesticide residues, if any, are present in different tropic levels in a Lake Michigan ecosystem' and 'to understand the biological significance, if any, of pesticide residues encountered in various layers of the Lake Michigan animal pyramid'. The bio-accumulation of DDT residues up the animal pyramid was clearly demonstrated and the findings are illustrated in Figure 15.4. Although bio-accumulation of DDT was known, the importance of the Lake Michigan study lay in the fact that it demonstrated that these problems could occur in a large lake system.

The second starting point was more dramatic. Michael Gilbertson, of the Canadian Wildlife Service, visited tern colonies on Lake Ontario in 1970. He described his visit as follows. 'As I walked about one of these islands, many birds whirled around my head and swooped down upon me again and again to prevent me from approaching their nests. Their shrill piercing cries rang in my ears. As I wandered about, I soon noticed that something was fundamentally wrong with the colony. While some young of varying age were found in the nests, the eggs in most had failed to hatch. On examining one of these eggs, I found that the young chick had died before it could completely crack open the shell. Several other eggs contained dead embryos. At the edge of a grass tussock, I also noticed an abnormal 2-week old chick, its upper and lower bill crossing over without meeting – a deformity which would result in certain starvation.'

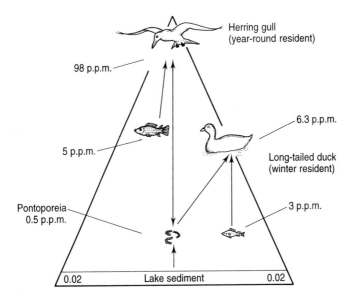

Figure 15.4 Accumulation of DDT up the Lake Michigan food chain. Values are total DDT on wet weight basis. After Hickey *et al.* (1966).

The herring gull became the key indicator species in the studies of pollution on the North American Great Lakes. The reasons for this choice are given in the Box 15.1.

Box 15.1 Reasons for the herring gull being a key indicator species on the Great Lakes

1. The herring gull feeds at the top of the food chains of the Great Lakes.

2. The adult herring gull is a year-round resident of the Great Lakes, with comparatively little movement from lake to lake.

3. The herring gull nests colonially. Colonial birds are probably the only groups of organisms for which it is possible to count the entire breeding population. Collection of eggs and assessment of reproduction are also easier in a colonial species.

4. The herring gull breeds throughout much of Europe and North America. This allows comparison between contaminant levels, reproductive success etc., on the Great Lakes with those in coastal and European populations.

Although the first studies, which had shown high embryonic mortality, were carried out in the mid-1960s, it was not until the mid-1970s that systematic studies were undertaken. The monitoring programme for residue levels was set up in 1974 with herring gull eggs being collected from colonies in all of the lakes and has continued until the present day. It was found that the reproductive success of herring gull was very low; in Lake Ontario only one young was produced for each 10 pairs nesting during the early 1970s.

The hypothesis that these severe effects over a wide area were caused by pollutants, notably the organochlorines, was immediately put forward. It was based on a correlation between total organochlorine levels and reproductive effects. A more difficult problem was relating these effects to a specific chemical or chemical(s). Apart from the question of possible interactions between chemicals, other issues arose. First, in areas where the level of one organochlorine was high, levels of other organochlorines were also high and vice versa. Second, the effects seen – death of embryos, structural abnormalities, behavioural changes – were known from laboratory studies to be caused by a wide range of organochlorines. Only in the case of eggshell thinning in cormorants (discussed later) was it possible to assign the cause to a specific chemical with a high degree of certainty.

One interesting study aimed to establish the relative importance of effects of pollutants in the adult birds which could cause behavioural changes (extrinsic effects) compared to effects of the pollutants in the egg of the developing embryo (intrinsic effects). In order to distinguish between these two causes of embryo mortality an egg exchange experiment was devised. The basis of this experiment was to move eggs from a highly contaminated (dirty) colony and place them under adults in a relatively uncontaminated (clean) colony and vice versa. Additionally, it is possible to incubate eggs from both 'clean' and 'dirty' colonies artificially to examine the effects of residues on the embryo. The outline of this experiment is shown in Table 15.1.

In theory this is a simple enough experiment; in practice it was not. Transportation did not affect the viability of the eggs, but to find colonies in different parts of the country with fresh eggs at the same time and move them rapidly from place to place was a logistical nightmare. Nevertheless, the results obtained from them in 1975 were quite clear: there was a major effect due to the pollutants in the egg and a significant effect of the pollutant on the behav-

Table 15.1 Outline of egg exchange experiments

Adult	Egg	Information obtained
Clean	Clean	Normal reproduction in clean environment
Clean	Dirty	Effect of pollutants in egg on the embryo
Dirty	Clean	Effect of pollutants on reproduction via behavioural changes of adult
Dirty	Dirty	Impaired reproduction in dirty environment

iour of the adult. We were foolish enough to repeat the experiment in 1976 when the results were less clear-cut, and again in 1977, when no differences were found at all. In fact, what we were seeing was a sudden marked improvement in the reproductive success of the herring gull in Lake Ontario.

While the increased breeding success of birds on the Great Lakes was good news, it did make the scientific investigation of the impact of pollutants very difficult. The cause of the change was the decreased inputs of some of the organochlorines into the environment. In 1972 the use of DDT was banned in the United States and its use curtailed in Canada. In the same year Monsanto introduced its voluntary ban on the 'open circuit uses' of PCBs. Thus, the intensive phase of the studies on the herring gull were carried out against a background of decreasing inputs.

The adverse effects on fish-eating birds were not confined to gulls. There was a severe decline of other species, such as the cormorant and the bald eagle. The best estimate of the population of cormorants on Lake Ontario in the 1940s and 1950s was 200 pairs; this had fallen to only three pairs by 1973. The main cause of failure was the breaking of eggshells. It was found at this time that 95% of the eggs broke before hatching. The recovery of the cormorant following the banning of DDT was rapid. By 1980 there were 375 pairs, and by 1990 no less than 6700 pairs. Over the last 5 years the population has been increasing at 25% per year; at this rate of increase the population would reach over 60 000 by the turn of the century and a million by 2010! Obviously, density-dependent factors – food supply, nesting habitat – will halt the increase before this point. An interesting question is why the population increased so much, way beyond the size of the initial population, once the pressure was released? The best explanation is a change in the fish stocks. Marked decreases in the population of the top predatory fish – lake trout and salmon – has allowed increases in the population of smaller fish such as alewife and smelt. The causes are complex, including both overfishing and the effects of pollutants on the reproduction of the larger, long-lived fish. However, when the reproduction improved as DDT levels decreased, the species were able to exploit the increased food base of small fish.

Abnormal young have been found in several species throughout the Great Lakes. The best information is available on the cormorant. In this case, the aid of bird watchers who ring these birds has been enlisted. They were asked to record the number of young with abnormalities, such as crossed beaks (Figure 15.1B), during the procedure of banding the young. The highest incidence of abnormalities was found in certain areas of Lake Michigan where one young in a hundred was defective. The rate of occurrence was considerably lower in other parts of the Great Lakes. Thus, although the occurrence of abnormalities was a potent reason for starting studies the actual impact of these on populations of fish-eating birds has been slight.

On the Great Lakes the scene had changed by the early 1980s. By this time the reproductive problems of fish-eating birds were now confined to a few specific areas, rather than occurring on a lake-wide basis as they had a decade before. On the other hand, the decline of levels of organochlorines which had been a feature of late 1970s has now halted and levels have remained essentially constant during the last half of the 1980s, so that now we are faced with a chronic pollution problem which can be expected to stretch into the next century.

Other changes had occurred since the early 1970s. In hindsight one problem that we had then was that neither the analytical chemistry or the experimental toxicology was sophisticated enough. The confirmation that the chlorinated dioxins (PCDDs) and chlorinated dibenzofurans (PCDFs) were present in the Great Lakes was not made until 1980. The chemistry of these compounds is discussed in section 1.2.2.

Nowadays analytical chemists routinely report on the levels of as many as 100 different organochlorines compared to a dozen a decade or so ago. Studies with specific congeners of PCBs, PCDFs and PCDDs have shown that the toxicity of specific compounds varies by several orders of magnitude. Toxicologists have demonstrated that 2,3,7,8-tetrachloro-p-dioxin is one of the most toxic compounds known but molecular biologists have found it useful for their work on the isolation and characterization of receptors. The identification of the Ah receptor in mid-1970s by means of its strong specific binding to dioxin was a key finding in bringing molecular biology into the realm of toxicology. The Ah receptor is responsible for the control of the monooxygenase enzyme systems which are responsible for an important defence mechanism against foreign compounds (see Chapters 7 and 9).

It was found that the ability of individual compounds to induce the monooxygenase system is greatly influenced by the degree of chlorination and the chlorine substitution pattern. The most toxic PCBs are those that have no chlorine atoms next to the central bond, which allows the molecule to rotate and fit into the receptor (see Chapter 1). Studies on the structure–activity relationship of a large number of organochlorines show that the toxic effects and strength of binding to the Ah receptor are related.

The application of this fundamental biochemistry to field investigations has involved expressing the complex mixtures of organochlorines as 'dioxin equivalents'. This concept is based on the correlation which has been found between the concentration required to induce a specific monooxygenase activity (alkyl hydrocarbon hydroxylase) and the concentration required for toxic effects for a large number of PCBs, PCDFs and PCDDs. In 'dioxin equivalents' the activity of the most powerful compound, 2,3,7,8-TCDD, is set at one and the potency of the other compounds are calculated from their ability to induce the monooxygenase. This is called the *toxic equivalent* of the compound. This number can then be multiplied by the concentration and the equivalence in terms of dioxin calculated. Although the potency of other compounds, such as individual PCBs, is much lower than dioxin, the concentrations are often much higher and therefore they frequently contribute more to the total 'dioxin equivalent' than dioxin itself. For example recent studies in Lake Michigan suggest that 90% of the dioxin equivalents in the eggs of fish-eating birds is

Table 15.2 Toxic equivalent factor and dioxin equivalent of three compounds

Compound	Concentration pg g^{-1}	Toxic equivalent factor	Dioxin equivalent
2,3,7,8-TCDD	2	1	2
3,3',4,4',5-PCB	3000	2×10^{-2}	60
3,3'4,4',5,5'-PCB	25 000	5×10^{-4}	12.5

caused by two specific PCBs. An example of the calculation of dioxin equivalents is given in Table 15.2. The approach is now used in reverse, the degree of induction of the enzyme is measured and then converted into dioxin equivalents. This bioassay approach is rapid and inexpensive compared to the conventional chemical analysis by gas chromatography – mass spectrometry.

The dioxin equivalents of egg samples from fish-eating birds collected from colonies of cormorants and terns in Michigan and Ontario have been determined. When the reproductive success of double-crested cormorant and Caspian tern (*Hydroprogne caspia*) colonies was plotted against dioxin equivalents of eggs from each colony a high degree of correlation was found (Figure 15.5).

The reproductive failure of fish-eating birds in the North American Great Lakes is an example where two initially entirely different lines of research – molecular biology and investigations by field biologists – eventually blended to provide an answer to the problem.

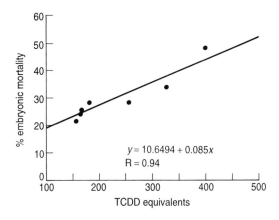

Figure 15.5 Relationship between dioxin equivalents and reproductive success in the Caspian tern on the North American Great Lakes. From Peakall (1992).

283

15.3 Reproductive Failure of Molluscs Caused by Tributyl tin

The tributyl tin (TBT) compounds have the general formula $(n-C_4H_9)_3-Sn-X$ where X is an anion such as chloride or carbonate. TBTs have been used as molluscicides on boats, quays and other marine structures and as biocides for cooling systems, in pulp and paper mills etc. Its use as an active ingredient in marine anti-fouling paints began in the mid-1960s, and for the next decade its popularity increased as it became widely recognized as being extremely effective. The worldwide production of organotins was estimated in 1980 to be about 30 000 tons annually; about a tenth of this was used in anti-fouling paints and a similar amount as wood perservatives.

Problems began to be recognized in the late 1960s with population declines of oysters and whelks in France and in Southern England and in marine snails in Long Island Sound in the United States. Investigations into the decline of populations of a mollusc, the dog whelk (*Nucella lapillus*), in Plymouth Sound showed that these declining populations exhibited a high degree of imposex. Imposex (females developing male characteristics) is typified by the development in females of a small penis close to the right tentacle. A superficial vas deferens grows between the genital papilla and the penis. In the most extreme cases this occludes the papilla thus preventing egg liberation and reproduction.

Detailed studies, both laboratory and field, have carried out by the Plymouth Marine Laboratory since the first finding of imposex in the dog whelk in 1969. A broader survey around the Southwest peninsula of England revealed that imposex was widespread with the most marked effects along the Channel coast.

One of the first findings was that there was a good relationship between the degree of imposex and the proximity of the affected population to harbours and marinas. This suggested a pollutant associated with the boating industry. Another suggestive finding was the marked increase of imposex after its discovery in 1969. These findings suggested the causative agent was TBT. These correlations were confirmed by laboratory experiments which allowed a cause-and-effect linkage to be made. These studies determined that a few nanograms per litre (ng l^{-1}) were enough to sterilize young whelks.

Imposex has been widely reported in marine gastropods associated with marinas and harbours around the world. Beside the studies already referred to in the UK, France and Eastern US, records include Western US and Canada, Alaska, Southeast Asia and New Zealand. Even more disturbing is the recent report of imposex in whelks from the open North Sea and its relationship to shipping traffic densities.

Laboratory experiments and in situ transfer experiments indicate that imposex may be initiated in dog whelks with concentrations of TBT as low as 1 ng l^{-1}. The no-observable-effect concentration (NOEC) for development of imposex is given as less than 1.5 ng l^{-1} in the International Programme on Chemical Safety Environmental Health Criteria Document published by the

World Health Organization in 1990. Sterilization of some females occurs at localities with 1–2 ng l^{-1} and is total in areas averaging 6–8 ng l^{-1}. Affected populations suffer reproductive failure and local extinction has occurred around marinas.

In France, restrictions on the use of TBT were initially brought in 1982 and have subsequently been extended to a complete ban of the use of organotin compounds in anti-fouling paints. In the UK, the use of TBT-containing anti-fouling paints on small boats and aquaculture cages was banned in 1987 and some recovery of populations of dog whelks has been reported since that time. In the United States a number of states have imposed restrictions or bans some of which are based on leaching rates. The leaching rate of TBT from paints has been considerably improved in recent years by incorporating the TBT in a co-polymer matrix.

15.3.1 *IPCS Environmental Health Criteria Documents*

Some 150 of these criteria documents have been published by the World Health Organization over the period 1962 to 1994. The document on TBT contains sections on identification and chemical properties, sources, transport and concentrations in the environment, effects on organisms, both field and experimental, and an evaluation of health risks to the environment and humans. Initially the document was prepared by two scientists and subsequently reviewed by a much larger expert panel.

15.4 Forest Spraying in Eastern Canada to Control Spruce Budworm

The spraying programme in Eastern Canada to control (or as some of its critics would say, to attempt to control) spruce budworm (*Choristoneura fumigerana*) has been the longest and largest programme of its kind in the world. In total, some 17 million kg of pesticides were sprayed on 67 million hectares of forest. The total area sprayed is approximately half the area of England and Wales, but some areas would have been sprayed annually or even twice annually. Even so, at the peak of the spraying operation (1974–1976) well over half of the province of New Brunswick was sprayed at least once annually (Figure 15.6). The operation is now on a much smaller scale, with 61 500 kg of fenitrothion sprayed on 293 000 hectares compared to 482 700 kg of fenitrothion and 56 200 kg of phosphamidon on 3 540 000 hectares in 1976. In the last decade the microbial agent *Bacillus thuringiensis* has been used operationally and accounted for 35% of the area sprayed in 1992. The total amounts of pesticides used in this programme are given in Table 15.3.

What have been the ecological consequences of this spraying? Impact studies showed that birds, especially canopy living birds, were the most vulnerable members of the fauna. Mammals, living on the forest floor, appeared

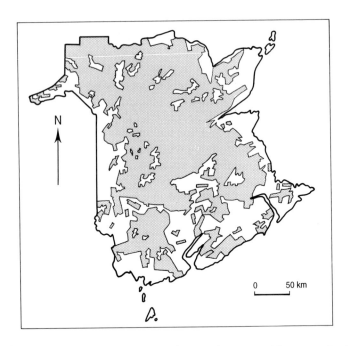

Figure 15.6 Map of New Brunswick, Canada, showing the extent of forest spraying (shaded) in 1976. From Pearce *et al.* (1976).

unaffected; amphibians and reptiles rather insensitive. Fish were severely affected by DDT, but not by organophosphorous or carbamate pesticides.

The impact of the organochlorine, DDT, was quite different from that of organophosphorus and carbamate pesticides which act by inhibiting cholinesterase. The most important, from a commercial point of view, was the mortality of young salmon. It was discovered that entire year classes of salmon had been eliminated in some important salmon rivers of maritime Canada. Less important commercially, but just as dramatic, was the loss from the area of the fish-eating hawk, the osprey. Also, the DDT used in forest spraying made a significant contribution to the DDT that caused the extinction of the peregrine

Table 15.3 Total amounts of pesticides (kg) used in New Brunswick in forest spray operations, 1952–1992

DDT (1952–64)	5 745 000
Phosphamidon (1963–77)	771 000
Fenitrothion (1969–92)	9 685 112
Aminocarb (1972–92)	551 762
Trichlorofon (1977–78)	289 000

falcon in Eastern North America, south of the arctic. This impact, mediated through reproductive failure caused by eggshell thinning, has already been discussed in some detail earlier in this chapter.

The first pesticide to replace DDT in the spray programmes in Eastern Canada was phosphamidon. It turned out to be a most effective avicide; the effect was quite different from DDT. DDT, accumulating through the food-chain, caused reproductive failure to those species that were sensitive to egg-shell thinning, but did not cause outright mortality at the dosages used in forest spraying operations. The amounts of DDT that caused widespread mortality of American robins following attempts to stop the spread of Dutch elm disease were much higher. In these programmes some 250–500 g of DDT was sprayed on individual trees.

The main tool for the calculation of the impact of pesticides on forest birds is the line transect of singing males. Repeated counts of singing male birds were made along roads and trails through the forest before and after spray treatment. The routes were 5 km long and took about 3 hours and were carried out in the early morning. This method was considered to be more effective than the intensive study of small plots, as in this latter method the numbers of birds recorded were often too small for statistical analysis to be valid. Another problem with the small-plot approach is that, in operational spraying, the coverage is far from uniform, so that a small plot may be under- or over-sprayed or even missed completely. The line transect, especially if undertaken at right-angles to the line of spray, does not have this weakness. Nevertheless, there are limitations to the approach, the most important of which are:

1. The counts may underestimate any decrease in vocal output as songs become more easily heard as their total number decreases.
2. The mortality of birds caused by the spray may be masked by immigration from unsprayed areas. The huge blocks of forest sprayed in Eastern Canada at the height of the spray programme make this less of a problem, although long-distance migration may occur.
3. The method does not prove that the cause of the change was the pesticide, although a marked decline immediately after the spray does strongly suggest that the pesticide was the cause. The most likely confounding factor is mortality caused by bad weather.
4. Even though a considerable number of transits are run and the results aver-aged, the degree of extrapolation involved is large.

It was estimated that nearly three million birds were killed in New Brunswick in 1975, due largely to phosphamidon. The impact of fenitrothion was con-sidered to be much less. These calculations, despite being open to criticism, do suggest that very considerable mortality occurred and were a major reason for the phasing out of phosphamidon in forest-spraying programmes.

Another useful tool is the determination of cholinesterase levels (see also Chapter 7). This technique has the advantage that inhibition of cholinesterase is clearly related to exposure to organophosphorus or carbamate pesticides. Further experimental studies have determined that chronic inhibition of 50% or acute inhibition of 80% are associated with mortality. Thus, a bird found dead with 80% inhibition of its brain cholinesterase can be diagnosed with certainty as have been killed by a spray operation. Nevertheless, under field conditions, there are several limitations to the approach. These limitations are:

1. The time course of AChE depression varies with different pesticides and probably varies with different species. It is often difficult to collect enough specimens of a given species at any one time interval after the spray operation.
2. Sampling may be influenced by birds coming into the area after spraying has occurred, so that their only exposure is secondary.
3. Birds with marked AChE inhibition are inactive, unwilling to fly or be flushed and are thus less likely to be captured or collected.

All these factors bias the data in favour of underestimating the degree of AChE inhibition. If the application rate is plotted against the degree of AChE inhibition (Figure 15.7), then one finds that the application rate known (from transient surveys or other population studies) to cause effects occurs at 35–40% inhibition (Mineau and Peakall, 1987).

It is clear that pesticides used in forest spraying can cause mortality, sometimes considerable mortality, at operational dosages. The dosage of fenitrothion commonly used in Canadian spray programmes (210 g ha^{-1}) does not have any appreciable safety margin. A considerable number of studies have

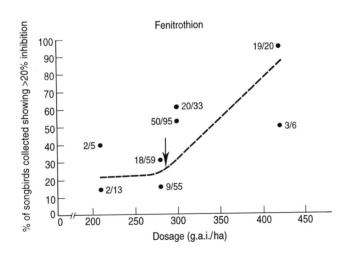

Figure 15.7 Relationship between inhibition of cholinesterase in songbirds and the dosage of fenitrothion applied to the forest. The arrow marks the dosage above which effects are seen on songbirds as judged by transect analysis.

Table 15.4 Surveys being run by the British Trust for Ornithology

Programme	Brief description	Information obtainable
Bird ringing	20 000 000 birds ringed since 1909	Movement of birds Longevity
Nest record Common bird Census	Started in 1939, now 30 000 nesting records annually	Reproductive success

Common bird
Census
	Started in 1961, 200 census
	    areas around the UK
		Population changes on 77
		    species

been carried out, some showing effects, others not. The safety margins for malathion and aminocarb are greater and several studies with these pesticides have not revealed adverse effects. Fenitrothion has been in use in New Brunswick for some 25 years now. The question may fairly be asked 'what are the effects of the persistent use of a non-persistent pesticide?'. Regrettably, the answer to that interesting question is that we do not know. It is, indeed, a difficult question to answer. First of all, the accurate measurement of long-term trends of avian (or indeed any wildlife) populations are difficult and expensive to determine. Even if this effort were made it is difficult, if not impossible, to assign a cause to the changes that have been established.

To collect enough information on populations it is necessary to have a large number of persons to carry out a large number of surveys for a long period of time. Virtually the only way to do this is to enlist amateurs (the word 'amateur' is used in its old sense of a person who does the work for the love of it, rather than any suggestion of the work being second-rate) as has been done by the British Trust for Ornithology. Some of the surveys currently being run by the BTO are given in Table 15.4. In addition, there are many more specialized surveys, including the heronies survey, garden bird survey, seabird colonies register, national wildfowl counts and the estuaries enquiry.

In North America the only survey of this type is the US Fish and Wildlife Services Breeding Bird Survey. This survey has shown declines of several species of canopy-dwelling passerines, but there are not enough data from the intensively sprayed areas to compare with other non-sprayed areas. Although the survey data do not give any direct indication of the cause of these declines, the most often cited cause is, in fact, destruction of wintering habitat in Latin America.

Further Reading

CADE T. J. *et al.* (eds) (1988) *Peregrine Falcon Populations: Their management and Recovery* A detailed account of the recovery of the peregrine and several other birds-of-prey throughout the world.

PEAKALL, D. B. (1993) A personal account of the discovery of eggshell thinning in the peregrine and its relationship to population changes.

ENVIRONMENT CANADA (1991) *Toxic Chemicals in Great Lakes and Associated Effects*, Vol. II. A detailed account (the first volume gives residue levels) of the changes documented in wildlife of the Great Lakes that have been attributed to toxic chemicals.

MINEAU, P. *et al.* (1984) A review of the residue levels in herring gull eggs and the biological changes that have been seen in this species over the period 1972–1980.

BRYAN, G. W. *et al.* (1986) The first major paper documenting the decline of the dog whelk and linkage of this decline to tributyl tin.

IPCS (INTERNATIONAL PROGRAMME ON CHEMICAL SAFETY). *Tributyl Tin Compounds*. A formal document that reviews the analytical methods, environmental levels and effects on organisms (ranging from microorganisms to man).

MINEAU, P. and PEAKALL, D. B. (1987) An evaluation of the methods – transects counts, number of singing male and cholinesterase levels – available to assess the impact of OP and carbamate pesticides used in forest spraying.

PEAKALL, D. B. and BART, J. R. (1983) A detailed review of the amounts of pesticides used, the areas sprayed and the effects seen in the spraying programmes in Eastern Canada and the Northeastern United States.

APPENDIX

Introduction to the Use of Population Projection Matrices (also known as Leslie Matrices)

A simple example will show how these matrices are used. Suppose the study organism lives for 3 years, and that:

the proportion of newborns that live to be 1 is 0.1
the proportion of 1-year-olds that live to be 2 is 0.8
the proportion of 2-year-olds that live to be 3 is 0.6
the proportion of 3-year-olds that live to be 4 is 0

Suppose the organism breeds at ages 1, 2 and 3, producing respectively 5, 10 and 3 offspring. These numbers are tabulated in a matrix, which we designate M1, as follows:

$$M1 = \begin{pmatrix} 0 & 5 & 10 & 3 \\ 0.1 & 0 & 0 & 0 \\ 0 & 0.8 & 0 & 0 \\ 0 & 0 & 0.6 & 0 \end{pmatrix}$$

This is an example of a 'population projection matrix'. Note that the age-specific fecundities are entered in order in the columns of the top row, and that the survivorships appear, in order, one per column, in the lower rows of the matrix. Thus the survivorship of the 0-year-olds appears in the second row in the first column, of 1-year-olds in the third row in the second column, and so on.

The reason for this peculiar method of tabulation is that it allows one to use matrix algebra, in the following way. Suppose that initially there are in the population 1000 0-year-olds, 50 1-year-olds, 50 2-year-olds and 10 3-year-olds. This is the initial *age-distribution* of the population, and is tabulated as a vector (column) that we shall refer to as C1, as follows:

$$C1 = \begin{pmatrix} 1000 \\ 50 \\ 50 \\ 10 \end{pmatrix}$$

The age distribution 1 year later, which we shall call C2, can now be obtained by matrix algebra:

C2 = M1 × C1

or, in full,

$$
C2 = \begin{pmatrix} 780 \\ 100 \\ 40 \\ 30 \end{pmatrix} = \begin{pmatrix} 0 & 5 & 10 & 3 \\ 0.1 & 0 & 0 & 0 \\ 0 & 0.8 & 0 & 0 \\ 0 & 0 & 0.6 & 0 \end{pmatrix} \times \begin{pmatrix} 1000 \\ 50 \\ 50 \\ 10 \end{pmatrix}
$$

Readers unfamiliar with matrix algebra might like to check that this gives the correct answer, i.e. that it calculates C2, the age distribution in year 2, correctly. Computers can be used to carry out the matrix algebra. Any computer software that performs matrix algebra may be used, e.g. Minitab. Furthermore the computer can be used to calculate the age distributions in later years, i.e. C3 = M1 × C2, C4 = M1 × C3 etc.

It will be found that after several years the population settles down and each age class then grows at the same rate each year. Specifically, each age class increases by a factor of 1.14 each year. Thus, the net reproductive rate, $\lambda = 1.14$. When all age classes increase by the same factor each year, a population is said to be in *stable age distribution*.

The population growth rate, $r = \log_e \lambda = 0.13$. When the population has reached its stable age distribution, each age class grows exponentially, with equation

$$
a_x(t) = a_x(0)e^{rt}
$$

where $a_x(t)$ represents the number of organisms in age class x, i.e. the number of age x at time t.

The value of λ eventually achieved does *not* depend on the initial age distribution. It satisfies the Euler–Lotka equation (equation 12.3 on page 216) and can be found by the iterative approach described there. An alternative method uses the fact that λ is the 'dominant eigenvalue' of the matrix M1. Dominant eigenvalues of matrices are easily found by appropriate computer software.

Glossary

AChE   Acetylcholinesterase.

acid rain   Rain made more acidic by the action of oxides of nitrogen and sulphur (pH below 5.5).

adducts   Products of the linkage between a xenobiotic and an endogenous molecule, e.g. adducts formed between benzo(a)pyrene metabolites and DNA.

ALAD   Amino laevulinic acid dehydrase.

allele   One of a pair of genes that occupy the same relative position on homologous chromosomes and separate during meiosis.

anion   A negatively charged atom or radical.

antagonism   Where the toxicity of a mixture is less than the sum of the toxicities of its components.

anthropogenic   Generated by the activities of man.

anthropogenic organic enrichment factor   Measurement of the contribution of human activity to the global cycle of a pollutant.

ATPases   Adenosine triphosphatases.

autotroph   An organism which obtains its carbon from carbon dioxide.

bio-accumulation factor (BAF) $\dfrac{\text{concentration of a chemical in an animal}}{\text{concentration of same chemical in its food}}$

bioconcentration factor (BCF) $\dfrac{\text{concentration of a chemical in an organism}}{\text{concentration of the same chemical in the ambient medium}}$

biomarker   A biological response to an environmental chemical which gives a measure of exposure and sometimes also of toxic effect. Usually restricted to responses at the level of organization of the whole organism or below.

293

biotransfomation Conversion of a chemical into one or more products by a biological mechanism (nearly always by enzyme action).

birth rate The number of offspring born to each reproductive female per year.

BOD (biochemical oxygen demand) The amount of dissolved oxygen used by micro-organisms to oxidize 1 litre of sewage sample.

carboxylesterases Esterases that hydrolyze organic compounds with carboxyester bonds. In the classification of enzymes by the International Union of Biochemistry, EC.3.1.1.1. refers to carboxylesterases which are inhibited by OP compounds.

carrying capacity (of the environment in which a population lives) The population size to which the population is driven by density-dependent processes. (See page 220.)

cation A positively charged atom or radical.

cetaceans Whales and dolphins.

CFC Chlorofluorocarbon.

ChE (cholinesterase) A general term for esterases which hydrolyze cholinesters.

CMPP 2-methyl-4-chloro-phenoxy propionic acid.

COD (chemical oxygen demand) The amount of oxygen required to achieve a complete chemical oxidation of 1 litre of a sewage sample.

congener A member of a group of structurally related compounds.

conjugate In biochemical toxicology, a molecule formed by the combination of a xenobiotic (usually a Phase I metabolite) with an endogenous molecule (e.g. glucuronic acid, glutathione, or sulphate).

contaminant See Introduction.

crankcase blowby Leakage around pistons into the crankcase of an internal-combustion engine.

curie A unit of measurement of radioactivity. 1 curie represents 3.7×10^{10} disinte-grations per second.

cytochrome P_{450} An iron-containing protein that catalyzes many biological oxida-tions.

2,4D 2,4-Dichlorophenoxyacetic acid.

2,4-DB 2,4-Dichlorophenoxybutyric acid.

2,4,5-T 2,4,5-Trichlorophenoxyacetic acid.

***pp*'DDT** *pp*'Dichlorodiphenyltrichloroethane.

density-dependence The phenomenon whereby factors vary in their effects with popu-lation density. (See page 220.)

DNA adduct See adduct.

DNOC Dinitroorthocresol.

e Base of natural logarithms (2.718).

EBI Ergosterol biosynthesis inhibitor (fungicide).

EC(D)$_{50}$ Concentration (dose) that affects designated criterion (e.g. behavioural trait) of 50% of a population. Also known as median effect concentration (dose).

endogenous Originating within an organism.

endoplasmic reticulum   Membranous network of cells which contains many enzymes that metabolize xenobiotics. Microsomes consist mainly of vesicles derived from endoplasmic reticulum.

epoxide hydrolase   Enzyme which converts epoxides to diols by the addition of water.

ester   An organic salt which will yield an acid and a base when hydrolyzed.

eukaryotes   Organisms that contain their DNA within nuclei.

eutrophication   The stimulation of algal growth in surface waters caused by high levels of nitrates and phosphates. Such pollution may be caused by sewage or run-off from agricultural land treated with fertilizers from industrial wastes.

eyrie   The nest of certain birds of prey, e.g. peregrine and golden eagle.

fitness   Used here to mean the population growth rate of an allele (note that this is different from the way fitness is defined in population genetics). (See page 242.)

fitness cost   The reduction in the fitness of resistant alleles relative to that of susceptible alleles in unpolluted environments. (See page 248.)

fugacity   A measurement of the 'escaping tendency' of a molecule (see Chapter 3).

GABA   Gamma amino butyric acid.

genotoxic   Toxic to the genetic material of the organism.

glucuronyl transferases   A group of enzymes that catalyze the formation of conjugates between glucuronic acid and a foreign compound (usually a Phase I metabolite).

Gray (Gy)   The SI unit of absorbed radiation dose corresponding to an energy absorption of 1 J kg^{-1} of matter (cf. Sievert).

haemprotein   A protein containing haeme as a prosthetic group, e.g. cytochrome P_{450}.

hazard   Potential of a chemical to cause harm.

hazard assessment   Comparison of ability to cause harm (see hazard) with expected environmental concentration.

HCB   Hexachlorobenzene.

hermaphrodite   An organism having both male and female characteristics.

heterotroph   An organism which obtains its carbon from organic compounds.

HMO   Hepatic microsomal monooxygenase (see Chapter 5).

immiscible   Refers to pairs of liquids which are unable to mix with one another, e.g. water and oil.

imposex   The imposition of male characters on females in prosobranch molluscs, principally the dog whelk, *Nucella lapillus*. A measure of imposex is the size of the female penis relative to that of the male. Within a population, imposex can be calculated with the following formula:

$$\text{level of imposex (\%)} = \frac{(\text{mean length of female penis mm})^3}{(\text{mean length of male penis mm})^3} \times 100$$

induction   With reference to enzymes: an increase in activity due to an increase in cellular concentration. This may be a response to a xenobiotic and usually involves an increased rate of synthesis of the enzyme.

k-value   A measure of mortality. (See pages 221–224.)

K Used generally as a biological constant; used in population ecology to mean carrying capacity.

K_{Ow} Octanol–water partition coefficient.

λ Net reproductive rate – a measure of the rate of increase of the population, viz. the factor by which population size is multiplied each year. $\lambda = e^r$. (See page 215.)

LC(D)$_{50}$ Concentration (dose) that kills 50% of the population observed. Also known as median lethal concentration (dose).

ligand A substance which binds specifically.

lipophilic Literally 'fat loving'. Lipophilic molecules have a high infinity for lipids and tend to move from water into membranes and fat depots.

lipoprotein An association between lipids and proteins.

logistic equation A simple equation showing how population growth rate may depend on population density. (See page 221.)

logistic regression A type of statistical regression that estimates the probability of an event occurring using one or more predictor variables.

macromolecule A large molecule, e.g. proteins, DNA and polysaccharides.

MCPA 2-methyl 4-chloro-phenoxyacetic acid.

melanism Possessing the dark pigment melanin.

metallothionein A metal-binding protein.

microcosm, mesocosm, macrocosm 'Small', 'medium' or 'large' multispecies system in which physical and biological parameters can be altered and subsequent effects monitored. They may be field- or laboratory-based and are thought to mimic responses of organisms in the field more realistically than single-species test systems.

microsomes/microsomal When tissue homogenates are subjected to differential centrifugation, the microsomal fraction is separated between approximately $10\,000\ g$ and $105\,000\ g$. It contains mainly vesicles derived from the endoplasmic reticulum in the case of most vertebrate tissues.

mitochondrion A subcellular organelle in which oxidative phosphorylation occurs, leading to the generation of ATP.

MO Monooxygenase. An enzyme system found in the endoplasmic reticulum of many cells. It contains cytochrome P_{450} as its active centre, and catalyzes the oxidation of many lipophilic organic compounds.

mutualism A relationship between two species in which each benefits from the presence of the other.

ozonosphere Layer of the atmosphere in which ozone is concentrated.

NOE(C)D No observed effect concentration or dose.

NOEL No observed effect level. See NOE(C)D.

OC Organochlorine compound.

OP Organophosphorus compound.

oxyradical An unstable form of oxygen containing an unpaired electron.

PAH Polycyclic aromatic hydrocarbon.

parthenogenetic Relating to 'virgin births'.

partition coefficient   See Chapter 5. K_{Ow} is an example of a partition coefficient.

PCB   Polychlorinated biphenyl.

PCDD   Polychlorinated dibenzodioxin.

PCDF   Polychlorinated dibenzofurans.

PEC   Predicted environmental concentration.

photochemical smog   A complex pollution arising as a consequence of photochemical reactions occurring in certain areas where there is strong solar radiation and high levels of aerial pollutants (e.g. from car exhaust fumes).

phytotoxic   Toxic to plants.

pinnipeds   Seals, sea-lions and walruses.

PNEC   Predicted no-effect concentration (see Chapter 6).

poikilotherms   Organisms that are unable to regulate their body temperatures.

polar   With reference to chemicals: molecules which have charge.

pollutant   See Introduction.

population growth rate, r   *Per capita* rate of increase of the population. (See page 215.)

porphyrins   Chemical structures which constitute prosthetic groups of certain proteins, e.g. cytochrome P_{450}, haemoglobin.

potentiation   With reference to toxicity: the situation where the toxicity of a combination of compounds is greater than the summation of the toxicities of its individual components.

probit analysis   A statistical procedure used to analyze data arising from toxicity tests. The response measured in the test is transformed into 'probit' values.

pyrethroids   Synthetic insecticides having a structural resemblance to the naturally occurring pyrethrins.

QSAR   Quantitative structure–activity relationships: relationships between parameters describing the structure of molecules and toxicity.

r, population growth rate (See page 215.)

rain-out   Removal of pollutants from air by incorporation into developing rain droplets of rain clouds.

reductase   An enzyme that performs reductions.

resistance   Reduced susceptibility to the toxicity of a chemical which is genetically determined (a characteristic of a resistant strain of animal or plant). (See pages 241–261.)

resistant allele   An allele which increases the fitness of its carriers in polluted environments. (See page 245.)

risk assessment   Probability of hazard being realized. Best expressed as fraction of a population (community) likely to be affected. In practice, however, hazard assessment is often carried out.

selective toxicity   A difference in the toxicity of a chemical between different species, sexes, strains or age groups. Expressed as a selectivity ratio, e.g.:

$$\frac{LD_{50} \text{ to species A}}{LD_{50} \text{ to species B}}$$

selectivity   See selective toxicity.

SFG   Scope for growth

SH   Sulphydryl group.

Sievert (Sv)   This SI unit is needed because the damage caused by radiation depends on the rate at which it is absorbed. Thus, a dose of relatively massive alpha particles of 20 Sv is typically equal to 1 Gray, whereas for the less damaging beta particles and gamma rays, typically 20 Sv = 20 Gy.

sister-chromatid exchange (SCE)   Reciprocal exchange of DNA at loci between chromatids during replication of DNA.

somatic growth rate   The rate of increase of an individual's body mass.

SR   Synergistic ratio – see Chapter 10.

standard deviation   A measure of the variation in a sample.

standard error   A measure of the precision of an estimate.

stereochemistry   A branch of chemistry which is concerned with the three-dimensional structure of chemicals.

sulphotransferases   Enzymes which catalyze the formation of conjugates between xenobiotics and sulphate.

survivorship   The proportion of animals surviving between two specified ages.

survivorship curve   Graph showing how survivorship from birth varies with age. (See page 218.)

susceptible allele   Non-resistant allele.

synergism   Similar to potentiation (q.v.), but some authors use the term in a more restricted way, e.g. where one component of a mixture (synergist) would cause no toxicity if applied alone at the stated dose.

TBT   Tributyl tin.

TCDD   Tetrachlorodibenzodioxin.

tolerance   The ability of an organism to withstand the adverse effects of pollution.

toxic   Harmful to living organisms.

toxicodynamics   Relating to the harmful effects of chemicals upon living organisms.

toxicokinetics   The uptake, metabolism, distribution and excretion of chemicals that express toxicity, i.e. concerned with the fate of chemicals in living organisms.

trade-off   Exchange of one advantageous character for another. (See page 244.)

vitellogenin   A protein that forms part of the yolk of egg-laying vertebrates.

wash-out   Removal of air pollutants by falling rain or snow.

xenobiotic   'Foreign compound'. A compound that is not part of the normal biochemistry of a defined organism. A foreign compound to one species may be a normal endogenous compound to another.

References

ÅHMAN, B. and ÅHMAN, G. (1994) Radiocesium in Swedish reindeer after the Chernobyl fallout: seasonal variations and long-term decline. *Health Physics* **66**, 503–512.

ALDRIDGE, W. N. (1953) Serum esterases I and II. *Biochemical Journal* **53**, 110–117, 117–124.

ALLAN, J. D. and DANIELS, R. E. (1981) Life table evaluation of chronic exposure to a pesticide. *Canadian Journal of Fisheries and Aquatic Sciences* **38**, 485–494.

ALLOWAY, B. J. and JACKSON, A. P. (1991) The behaviour of heavy metals in sewage sludge-amended soils. *Science of the Total Environment* **100**, 151–176.

ANDREWARTHA, H. G. and BIRCH, L. C. (1954) *The Distribution and Abundance of Animals.* Chicago: University of Chicago Press.

ATCHISON, G. J., SANDHEINRICH, M. B. and BRYAN, M. D. (1996) In NEWMAN, M. C. and JAGOE, C. H. (eds) *Quantitative Ecotoxicology: a Hierarchical Approach.* Chelsea, MI: Lewis.

BACCI, E. (1993) *Ecotoxicology of Organic Contaminants.* Boca Raton, FL: Lewis.

BAKER, A. J. M. (1987) Metal tolerance. *New Phytologist* **106** (supplement), 93–111.

BAKER, A. J. M. and PROCTOR, J. (1990) The influence of cadmium, copper, lead and zinc on the distribution and evolution of metallophytes in the British Isles. *Plant Systematics and Evolution* **173**, 91–108.

BAKER, A. J. M. and WALKER, P. L. (1989) Physiological responses of plants to heavy metals and the quantification of tolerance and toxicity. *Chemical Speciation and Bioavailability* **1**, 7–12.

BALK, F. and KOEMAN, J. H. (1984) Future hazards from pesticide use. Commission on Ecology Papers No. 6. International Union for the Conservation of Nature and Natural Resources.

BALLANTYNE, B. and MARRS, T. C. (1992) *Clinical Experimental Toxicology of Organophosphates and Carbamates.* Oxford: Butterworth/Heinemann.

BARTH, H. (ed.) (1987) *Reversibility of Acidification.* London: Elsevier.

BEEBY, A. (1991) Toxic metal uptake and essential metal regulation in terrestrial invertebrates: a review. In NEWMAN, M. C. and MCINTOSH (eds) *Metal Eco-*

toxicology: Concepts and Applications, pp. 65–89. Boca Raton, MI: Lewis.

BEGON, M., MORTIMER, M and THOMPSON, D.J. (1996) *Population Ecology: a Unified Study of Animals and Plants*. Oxford: Blackwell Scientific Publications. (3rd edition).

BEITINGER, T. L. (1990) Behavioral reactions for the assessment of stress in fishes. *Journal of Great Lakes Research* **16**, 495–528.

BENGTSSON, G., GUNNARSSON, T. and RUNDGREN, S. (1983) Growth changes caused by metal uptake in a population of *Onychiurus armatus* (Collembola) feeding on metal polluted fungi. *Oikos* **40**, 216–225.

BENGTSSON, G., GUNNARSSON, T. and RUNDGREN, S. (1985) Influence of metals on reproduction, mortality and population growth in *Onychiurus armatus* (Collembola). *Journal of Applied Ecology* **22**, 967–978.

BENN, F. R. and MCAULIFFE, C. A. (eds) (1975) *Chemistry and Pollution*. London: Macmillan.

BERGGREN, D., BERGVIST, B., FALKENGREN-GRERUP, U., FOLKESON, L. and TYLER, G. (1990) Metal solubility and pathways in acidified forest ecosystems of South Sweden. *Science of the Total Environment* **96**, 103–114.

BISHOP, J. A. (1981) A neoDarwinian approach to resistance: examples from mammals. In BISHOP, J. A. and COOK, L. M. (eds) *Genetic Consequences of Man-made Change*, pp. 37–51. London: Academic Press.

BISHOP, J. A. and COOK, L. M. (1981) *The Genetic Consequences of Man-made Change*. London: Academic Press.

BOON, J. P., EIJGENRAAM, F., EVERAARTS, J. M. and DUINKER, J. C. (1989) A structure–activity relationship (SAR) approach towards metabolism of PCBs in marine animals from different trophic levels. *Marine Environmental Research* **27**, 159–176.

BORG, H., ANDERSSON, P. and JOHANSSON, K. (1989) Influence of acidification on metal fluxes in Swedish forest lakes. *Science of the Total Environment* **87/88**, 241–253.

BORIO, R., CHIOCCHINI, S., CICIONI, R., ESPOSTI, P. D., RONGONI, A., SABATINI, P., SCAMPOLI, P., ANTONINI, A. and SALVADORI, P. (1991) Uptake of radiocesium by mushrooms. *Science of the Total Environment* **106**, 183–190.

BOSVELD, A. T. C. and VAN DEN BERG, M. (1994) Effects of polychlorinated biphenyls, dibenzo-*p*-dioxins and dibenzofurans on fish-eating birds. *Environmental Reviews* **2**, 147–166.

BOWER, J. S., BROUGHTON, G. F. J., STEDMAN, J. R. and WILLIAMS, M. L. (1994) A winter NO_2 smog episode in the UK. *Atmospheric Environment* **28**, 461–475.

BRAKEFIELD, P. M. (1987) Industrial melanism: do we have the answers? *Trends in Ecology and Evolution* **2**, 117–122.

BRITISH ECOLOGICAL SOCIETY (1990) River Water Quality. Ecological Issues Number 1, Field Studies Council, Preston Montford.

BROOKS, G. T. (1974) *Chlorinated Insecticides* (2 vols). Cleveland, OH: CRC Press.

BROUWER, A. and VAN DEN BERG, K. J. (1986) Binding of a Metabolite of 3,4,3', 4'-tetrachlorobiphenyl to transthyretin reduces serum vitamin A transport by inhibiting the formation of the protein complex carrying both retinol and thyroxin. *Toxicology and Applied Pharmacology* **85**, 301–312.

BROUWER, A., REIJNDERS, P. J. H. and KOEMAN, J. H. (1989) Polychlorinated biphenyl (PCB)-contaminated fish induces vitamin A and thyroid hormone deficiency in the common seal (*Phoca vitulina*). *Aquatic Toxicology* **15**, 99–105.

BROUWER, A., MURK, A. J. and KOEMAN, J. H. (1990) Biochemical and physiological approaches in ecotoxicology. *Functional Ecology* **4**, 275–281.

BRYAN, G. W. and LANGSTON, W. J. (1992) Bioavailability, accumulation and effects of heavy metals in sediments with special reference to United Kingdom estuaries: a review. *Environmental Pollution* **76**, 89–131.

BRYAN, G. W., GIBBS, P. E., HUMMERSTONE, L. G. and BURT, G. R. (1986) The decline of the gastropod *Nucella lapillus* around South-West England: evidence for the effect of tributyltin from antifouling paints. *Journal of the Marine Biological Association of the United Kingdom* **66**, 611–640.

BRYAN, G. W., GIBBS, P. E. and BURT, G. R. (1988) A comparison of the effectiveness of tri-n-butyltin chloride and five other organotin compounds in promoting the development of imposex in the dog whelk *Nucella lapillus*. *Journal of the Marine Biological Association of the United Kingdom* **68**, 733–744.

BUNCE, N. (1991) *Environmental Chemistry*. Winnipeg: Wuerz.

BUSBY, D. G., WHITE, L. M. and PEARCE, P. A. (1990) Effects of aerial spraying of fenitrothion on breeding white-throated sparrows. *Journal of Applied Ecology* **27**, 743–755.

BUTLER, J. D. (1979) *Air Pollution Chemistry*. London: Academic Press.

CADE, T. J., ENDERSON, J. H., THELANDER, C. G. and WHITE, C. M. (eds) (1988) *Peregrine Populations: Their Management and Recovery*. Boise, ID: The Peregrine Fund Inc.

CALAMARI, D. and VIGHI, M. F. (1992) Role of evaluative models to assess exposure to pesticides. In TARDIFF, R. G. (ed.) *Methods to Assess Adverse Organisms (SCOPE)*, pp. 119–132. Chichester: John Wiley.

CALLAGHAN, C. A., MENZIE, C. A., BURMASTER, D. E., WILBORN, D. C. and ERNST, T. (1991) On-site methods for assessing chemical impact on the soil environment using earthworms: a case study at the Baird and McGuire superfund site. *Environmental Toxicology and Chemistry* **10**, 812–826.

CALOW, P. (1989) Ecotoxicology? *Journal of Zoology* **218**, 701–704.

CALOW, P. (ed.) (1993) *Handbook of Ecotoxicology*, vol 1. Oxford: Blackwell.

CALOW, P. (ed.) (1994) *Handbook of Ecotoxicology*, vol 2. Oxford: Blackwell.

CARTER, L. J. (1976) Michigan's PBB incident: chemical mix-up leads to disaster. *Science* **192**, 240–243.

CASWELL, H. (1989) *Matrix Population Models*. Sunderland, MA: Sinauer Associates Inc.

CLARKE, B. (1975) The contribution of ecological genetics to evolutionary theory: detecting the direct effects of natural selection on particular polymorphic loci. *Genetics* **79**, 101–108.

CLARKE, R. B. (1992) *Marine Pollution*, 3rd edn. Oxford: Clarendon Press.

COHEN, and RINGER, D. (1975) Lead. Possible toxicity in urban vs rural rats. *Archives Environmental Health* **30**, 276–280.

COLEMAN, J. E. (1967) Metal ion dependent binding of sulphonamide to carbonic anhydrase. *Nature* **214**, 193–194.

CROMMENTUIJN, T., BRILS, J. and VAN STRAALEN, N. M. (1993) Influence of cadmium on life-history characteristics of *Folsomia candida* (Willem) in an artificial soil substrate. *Ecotoxicology and Environmental Safety* **26**, 216–227.

CROSSLAND, N. O. (1988) A method for evaluating effects of toxic chemicals on fish growth rates. In ADAMS, J. A., CHAPMAN, G. A. and LANDIS, W. G. (eds) *Aquatic Toxicology and Hazard Assessment*, 10th vol. Philadelphia: American

Society for Testing and Materials.

CROSSLAND, N. O. (1994) Extrapolating from mesocosms to the real world. *Toxicology and Ecotoxicology News* 1, 15–22.

CROUT, N. M. J., BERESFORD, N. A. and HOWARD, B. J. (1991) The radiological consequences for lowland pastures used to fatten upland sheep contaminated with radiocaesium. *Science of the Total Environment* 103, 73–88.

CULBARD, E. B., THORNTON, I., WATT, J., WHEATLEY, M., MOORCROFT, S. and THOMPSON, M. (1988). Metal contamination in British urban dusts and soils. *Journal of Environmental Quality* 17, 226–234.

CURTIS, C. F., COOK, L. M. and WOOD, R. J. (1978) Selection for and against insecticide resistance and possible methods of inhibiting the evolution of resistance in mosquitoes. *Ecological Entomology* 3, 273–287.

DALLINGER, R. (1993) Strategies of metal detoxification in terrestrial invertebrates. In DALLINGER, R. and RAINBOW, P. S. (eds) *Ecotoxicology of Metals in Invertebrates*, pp. 245–289. Chelsea, USA: Lewis.

DALLINGER, R. and RAINBOW, P. S. (eds) (1993) *Ecotoxicology of Metals in Invertebrates*. Chelsea, USA: Lewis.

DANIELS, R. E. and ALLAN, S. D. (1981) Life table evaluation of chronic exposure to pesticide. *Canadian Journal of Fisheries and Aquatic Sciences* 38, 485–494.

DEININGER, R. A. (1987) Survival of Father Rhine. *Journal of American Water Works Association* 79, 78–93.

DEMPSTER, J. P. (1975) Effects of organochlorine insecticides on animal populations. In MORIARTY, F. (ed.) *Organochlorine Insecticides: Persistent Organic Pollutants*, pp. 231–248. London: Academic Press.

DEPLEDGE, M. H. (1994) The rational basis for the use of biomarkers as ecotoxicological tools. In FOSSI, C. M. and LEONZIO, C. (eds) *Nondestructive Biomarkers in Vertebrates*, pp. 271–295. Boca Raton, FL: Lewis.

DEPLEDGE, M. H., AMARAL-MENDEL, J. J., DANIEL, B., HALBROOK, R. S., KLOEPPER-SAMS, P., MOORE, M. N. and PEAKALL, D. (1993) The conceptual basis of the biomarker approach. In PEAKALL, D. B. and SHUGART, R. L. (eds) *Biomarkers. Research and Application in the Assessment of Environmental Health*, pp. 15–29. Berlin: Springer-Verlag.

DEVONSHIRE, A. L. and SAWICKI, R. M. (1979) Insecticide-resistant *Myzus persicae* as an example of evolution by gene duplication. *Nature* 280, 140–141.

DIX, H. M. (1981) *Environmental Pollution: Atmosphere, Land, Water and Noise*. Chichester: John Wiley.

DONKER, M. H., EIJSACKERS, H. and HEIMBACH, F. (eds) (1994) *Ecotoxicology of Soil Organisms*. Boca Raton, FL: Lewis.

DROBNE, D. and HOPKIN, S. P. (1994) Ecotoxicological laboratory test for assessing the effects of chemicals on terrestrial isopods. *Bulletin of Environmental Contamination and Toxicology* 53, 390–397.

EDWARDS, C. A. (1976) *Persistent Pesticides in the Environment*, 2nd edn. Boca Raton, FL: CRC Press.

EDWARDS, T. (1994) Chornobyl [sic!]. *National Geographic* 186, 100–115

ELLIOTT, J. M. (1993) A 25-year study of production of juvenile Sea-trout, *Salmo trutta*, in an English Lake District stream. *Canadian Special Publication of Fisheries and Aquatic Sciences* 118, 109–122.

ENDERSON, J. H., TEMPLE, S. A. and SWARTZ, L. G. (1972) Time-lapse photographic records of nesting Peregrine Falcons. *Living Bird* 11, 113–128.

References

ENVIRONMENT CANADA (1991) *Toxic Chemicals in Great Lakes and Associated Effects*, vol II, *Effects*, pp. 495–755. Toronto: Environment Canada.

ENVIRONMENTAL HEALTH CRITERIA NO. 6 (1989) Mercury-Environmental Aspects. Geneva: WHO.

ENVIRONMENTAL HEALTH CRITERIA NO. 63 (1986) Organophosphorus Insecticides – a General Introduction. Geneva: WHO.

ENVIRONMENTAL HEALTH CRITERIA NO. 64 (1986) Carbamate Pesticides – a General Introduction. Geneva: WHO.

ENVIRONMENTAL HEALTH CRITERIA NO. 101 (1990) Methylmercury. Geneva: WHO.

ENVIRONMENTAL HEALTH CRITERIA NO. 116 (1990) Tributyl Tin compounds. Geneva: WHO.

ERNST, W. (1976) Physiological and biochemical aspects of metal tolerance. In MANSFIELD, T. A. (ed.) *Effects of Air Pollutants on Plants*. Cambridge: Cambridge University Press.

ERNST, W. H. O. and PETERSON, P. J. (1994) The role of biomarkers in environmental assessment. (4) Terrestrial plants. *Ecotoxicology* 3, 180–192.

ERNST, W. H. O., VERKLEIJ, J. A. C. and SCHAT, H. (1992) Metal tolerance in plants. *Acta Botanica Neerlandica* 41, 229–248.

ETO, M. (1974) *Organophosphorus Insecticides: Organic and Biological Chemistry*. Cleveland, OH: CRC Press.

EVANS, P. R. (1990) Population Dynamics in relation to pesticide use, with particular reference to birds and mammals. In SOMERVILLE, L. and WALKER, C. H. (eds) *Pesticide Effects on Terrestrial Wildlife*, pp. 307–317. London: Taylor and Francis.

EVERAARTS, J. M., SHUGART, L. R., GUSTIN, M. K., HAWKINS, W. E. and WALKER, W. W. (1993) Biological markers in fish: DNA integrity, hematological parameters and liver somatic index. *Marine Environmental Research* 35, 101–107.

FAIRBROTHER, A. (1994) Clinical biochemistry. In FOSSI, M. C. and LEONZIO, C. (eds) *Nondestructive Biomarkers in Vertebrates*, pp. 63–89. Boca Raton, FL: Lewis.

FAIRBROTHER, A., MARDEN, B. T., BENNETT, J. K. and HOOPER, M. J. (1991) Methods used in determination of cholinesterase activity. In MINEAU, P. (ed.) *Cholinesterase-Inhibiting Insecticides – their impact on Wildlife and the Environment*, pp. 35–71. Amsterdam: Elsevier.

FEDER, M. E., BENNETT, A. F., BURGGREN, W. W. and HUEY, R. B. (1987) *New Directions in Ecological Physiology*. Cambridge: Cambridge University Press.

FEST, C. and SCHMIDT, K.-J. (1982) *Chemistry of Organophosphorus Pesticides*, 2nd edn. Berlin: Springer-Verlag.

FISCHER, H. (1989) Cadmium in seawater recorded by mussels: regional decline established. *Marine Ecology Progress Series* 55, 159–169.

FORRESTER, N. W., CAHILL, M., BIRD, L. J. and LAYLAND, J. K. (1993) Management of pyrethroid and endosulfan resistance in *Helicoverpa armigera* (Lepidoptera: Noctuidae) in Australia. *Bulletin of Entomological Research* (supplement 1), 1–132.

FOSSI, M. C. and LEONZIO, C. (eds) (1993) *Nondestructive Biomarkers in Vertebrates*. Boca Raton, FL: Lewis.

FOX, G. A., KENNEDY, S. W., NORSTROM, R. J. and WINGFIELD, D. C. (1988) Porphyria in herring gulls: a biochemical response to chemical contamination of Great Lake food chains. *Environmental Toxicology and Chemistry* 7, 831–839.

FRANK, P. W., BOLL, C. D. and KELLY, R. W. (1957) Vital statistics of laboratory cultures of *Daphnia pulex* DeGeer as related to density. *Physiological Zoology* **30**, 287–305.

FREEDMAN, B. (1989) *Environmental Ecology: the Impacts of Pollution and Other Stresses on Ecosystem Structure and Function.* San Diego: Academic Press.

FREEDMAN, B. (1995) *Environmental Ecology: the Environmental Effects of Pollution Disturbance and Other Stresses*, 2nd edn. San Diego: Academic Press.

FRYDAY, S. L., HART, A. D. M. and DENNIS, N. J. (1994) Effects of exposure to an organophosphate on the seed-handling efficiency of the House sparrow. *Bulletin of Environmental Contamination and Toxicology* **53**, 869–876.

FUTUYMA, D. J. (1986) *Evolutionary Biology*, 2nd edn. Sunderland, MA: Sinauer Associates.

GIBBS, P. E. (1993) A male genital defect in the dog-whelk, *Nucella lapillus* (NEOGASTROPODA), favouring survival in a TBT-polluted area. *Journal of the Marine Biological Association of the United Kingdom* **73**, 667–678.

GIBBS, P. E. and BRYAN, G. W. (1986) Reproductive failure in populations of the dog whelk *Nucella lapillus*, caused by imposex induced by tributyltin from antifouling paints. *Journal of the Marine Biological Association of the United Kingdom* **66**, 767–777.

GIBBS, P. E., PASCOE, P. L. and BURT, G. R. (1988) Sex change in the female dog whelk, *Nucella lapillus*, induced by tributyl tin from antifouling paints. *Journal of the Marine Biological Association of the United Kingdom* **68**, 715–731.

GIBSON, G. and SKETT, P. (1986) *Introduction to Drug Metabolism.* London: Chapman and Hall.

GOUGH, J. J., McINDOE, E. C. and LEWIS, G. B. (1994) The use of dimethoate as a reference compound in laboratory acute toxicity tests on honey bees (*Apis mellifera* L.) 1981–1992. *Journal of Apicultural Research* **33**, 119–125.

GRANT, A. and MIDDLETON, R. (1990) An assessment of metal contamination in the Humber Estuary, U.K. *Estuarine, Coastal and Shelf Science* **31**, 71–85.

GRAY, J. S. (1981) *The Ecology of Marine Sediments.* Cambridge: Cambridge University Press, 185 pp.

GREIG-SMITH, P. W. and HARDY, A. R. (1992) Design and management of the Boxworth project. In GREIG-SMITH, P., FRAMPTON, G. and HARDY, T. (eds) *Pesticides, Cereal Farming and the Environment. The Boxworth Project.* London: HMSO.

GREIG-SMITH, P. W., BECKER, H., EDWARDS, P. J. and HEIMBACH, F. (eds) (1992) *Ecotoxicology of Earthworms.* Andover: Intercept.

GREIG-SMITH, P. W., FRAMPTON, G. and HARDY, T. (1992) *Pesticides, Cereal Farming and the Environment. The Boxworth Project.* London: HMSO.

GRUE, C. E., FLEMING, W. J., BUSBY, D. G. and HILL, E. F. (1983) Assessing hazards of organophosphate pesticides to wildlife. *Transactions North American Wildlife Conference* **48**, 200–220.

GRUE, C. E., HOFFMAN, D. J., BEYER, W. N. and FRANSON, L. P. (1986) Lead concentrations and reproductive success in European starlings *Sturnus vulgaris* nesting within highway roadside verges. *Environmental Pollution* **42A**, 157–182.

GRUE, C. E., HART, A. D. M. and MINEAU, P. (1991) Biological consequences of depressed brain cholinesterase activity in wildlife. In MINEAU, P. (ed.) *Cholinesterase-Inhibiting Insecticides – Their Impact on Wildlife and the Environment*, pp. 151–210. Amsterdam: Elsevier.

GUTHRIE, F. E. and PERRY, J. J. (eds) (1980) *Introduction to Environmental Toxi-*

cology. New York: Elsevier, 484 pp.

HÅGVAR, S. and ABRAHAMSEN, G. (1990) Microarthropoda and Enchytraeidae (Oligochaeta) in naturally lead-contaminated soil: a gradient study. *Environmental Entomology* **19**, 1263–1277.

HAMER, D. M. (1986) Metallothionein. *Annual Review of Biochemistry* **55**, 913–951.

HART, A. D. M. (1993) Relationships between behaviour and the inhibition of acetylcholinesterase in birds exposed to organophosphorus pesticides. *Environmental Toxicology and Chemistry* **12**, 321–336.

HART, I. G., SHULTICE, R. W. and FOUTS, J. R. (1963) Stimulatory effects of chlordane on hepatic microsomal drug metabolism in the rat. *Toxicology and Applied Pharmacology* **5**, 371–386.

HARTL, D. L. and CLARK, A. G. (1989) *Principles of Population Genetics.* Sunderland, MA: Sinauer Associates.

HARTWELL, S. I., CHERRY, D. S. and CAIRNS, J. JR (1987) Field validation of avoidance of elevated metals by fathead minnows (*Pimephales promelas*) following *in situ* acclimation. *Environmental Toxicology and Chemistry* **6**, 189–200.

HASSALL, K. A. (1990) *The Biochemistry and Uses of Pesticides*, 2nd edn. London: Macmillan.

HEDGECOTT, S. (1994) Prioritization and standards for hazardous chemicals. In CALOW, P. (ed.) *Handbook of Ecotoxicology*, pp. 368–393, Oxford: Blackwell.

HEGDAL, P. L. and BLASKIEWICZ, R. W. (1984) Evaluation of the potential hazard to Barn Owls of talon (brodifacoum bait) used to control rats and house mice. *Environmental Toxicology and Chemistry* **3**, 167–179.

HEGGESTAD, H. E. (1991) Origin of Bel-W3, Bel-C and Bel-B tobacco varieties and their use as indicators of ozone. *Environmental Pollution* **74**, 264–291.

HELIÖVAARA, K., VÄISÄNEN, R., BRAUNSCHWEILER, H. and LODENIUS, M. (1987) Heavy metal levels in two biennial pine insects with sap-sucking and gall-forming life styles. *Environmental Pollution* **48**, 13–23.

HICKEY, J. J. (1969) *The Peregrine Falcon Populations: Their Biology and Decline.* Madison, WI: University of Wisconsin Press.

HICKEY, J. J., KEITH, J. A. and COON, F. B. (1966) An exploration of pesticides in a Lake Michigan ecosystem. *Journal of Applied Ecology* **3** (suppl), 141–154.

HILL, E. F. and FLEMING, W. J. (1982) Anticholinesterase poisoning of birds: field monitoring and diagnosis of acute poisoning. *Environmental Toxicology and Chemistry* **1**, 27–38.

HODGSON, E. and KUHR, R. J. (eds) (1990) *Safer Insecticides – Development and Use.* New York: M. Dekker.

HODGSON, E. and LEVI, P. (1993) *Introduction to Biochemical Toxicology*, 2nd edn. Norwalk, CT: Appleton and Lange.

HOFFMAN, A. A. and PARSONS, P. A. (1991) *Evolutionary Genetics and Environmental Stress.* Oxford: University Press.

HOLLOWAY, G. J., SIBLY, R. M. and POVEY, S. R. (1990) Evolution in toxin-stressed environments. *Functional Ecology* **4**, 289–294.

HOPKIN, S. P. (1989) *Ecophysiology of Metals in Terrestrial Invertebrates.* Barking (UK): Elsevier Applied Science.

HOPKIN, S. P. (1990) Critical concentrations, pathways of detoxification and cellular ecotoxicology of metals in terrestrial arthropods. *Functional Ecology* **4**, 321–327.

HOPKIN, S. P. (1993a) *In situ* biological monitoring of pollution in terrestrial and aquatic ecosystems. In CALOW, P. (ed.) *Handbook of Ecotoxicology*, vol 1, pp.

397–427. Oxford: Blackwell.

HOPKIN, S. P. (1993b) Deficiency and excess of copper in terrestrial isopods. In DALLINGER, R. and RAINBOW, P. S. (eds) *Ecotoxicology of Metals in Invertebrates*, pp. 359–382. Chelsea, USA: Lewis.

HOPKIN, S. P. (1993c) Ecological implications of '95% protection levels' for metals in soils. *Oikos* **66**, 137–141.

HOPKIN, S. P. (1995) Deficiency and excess of essential and non-essential metals in terrestrial insects. *Symposia of the Royal Entomological Society of London* **17**, 251–270.

HOPKIN, S. P. and MARTIN, M. H. (1984) Heavy metals in woodlice. *Symposia of the Zoological Society of London* **53**, 143–166.

HOPKIN, S. P., HARDISTY, G. N. and MARTIN, M. H. (1986) The woodlouse *Porcellio scaber* as a 'biological indicator' of zinc, cadmium, lead and copper pollution. *Environmental Pollution* (Series B) **11**, 271–290.

HOPKIN, S. P., HAMES, C. A. C. and DRAY, A. (1989) X-ray microanalytical mapping of the intracellular distribution of pollutant metals. *Microscopy and Analysis* **14**, 23–27.

HOVE, K., PEDERSEN, O., GARMO, T. H., HANSEN, H. S. and STAALAND, H. (1990) Fungi: a major source of radiocesium contamination of grazing ruminants in Norway. *Health Physics* **59**, 189–192.

HUCKLE, K. R., WARBURTON, P. A., FORBES, S. and LOGAN, C. J. (1989) Studies on the fate of flucoumafen in the Japanese Quail (*Coturnix coturnix japonica*). *Xenobiotica* **18**, 51–62.

HUDSON, R. H., TUCKER, R. K. and HAEGELE, M. A. (1984) *Handbook of Toxicity of Pesticides to Wildlife*. US Fish and Wildlife Service Resource Publishers 153, 90 pp.

HUGGETT, R. J., KIMERLE, R. A., MEHRLE, P. M. JR. and BERGMAN, H. L. (eds) (1992) *Biomarkers. Biochemical, Physiological, and Histological Markers of Anthropogenic Stress*. Boca Raton, FL: Lewis.

HUTTON, M. (1980) Metal Contamination of feral Pigeons *Columba livia* from the London area: Part 2 – Biological effects of lead exposure. *Environmental Pollution* **22A**, 281–293.

HYNES, H. B. N. (1960) *The Biology of Polluted Waters*. Liverpool: Liverpool University Press.

INOUE, K., OSAKABE, M., ASHIHARD, W. and HAMAMURA, T. (1986) Factors affecting abundance of the Kanzawa spider mite *Tetranychus kanzawai* on grapevine in a glasshouse: influence of pesticidal application on occurrence of the Kanzawa spider mite and its predators. *Bulletin of the Fruit Tree Reseach Station*, Series E, 103–116.

IPCS (INTERNATIONAL PROGRAMME ON CHEMICAL SAFETY) ENVIRONMENTAL HEALTH CRITERIA No. 116 (1990) Tributyl tin compounds. Geneva: WHO.

JOHNSTON, G. O. (1995) The study of interactive effects of pesticides in birds – a biomarker approach. *Aspects of Applied Biology* **41**, 25–31.

JONES, K. C., SYMON, C., TAYLOR, P. J. L., WALSH, J. and JOHNSTON, A. E. (1991) Evidence for a decline in rural herbage lead levels in the U. K. *Atmospheric Environment* **25A**, 361–369.

JORGENSEN, S. E. (ed.) (1991) *Modelling in Ecotoxicology*. Amsterdam: Elsevier, 356 pp.

JUKES, T. (1985) Selenium not for dumping. *Nature* **349**, 438–440.

KAZAKOV, V. S., DEMIDCHIK, E. P. and ASTAKHOVA, L. N. (1992) Thyroid cancer after Chernobyl. *Nature* **359**, 21.

KENNEDY, S. W. and JAMES, C. A. (1993) Improved method to extract and concentrate porphyrins from liver tissue for analysis by high-performance liquid chromatography. *Journal of Chromatography* **619**, 127–132.

KETTLEWELL, B. (1973) *The Evolution of Melanism*. Oxford: Clarendon Press.

KOEMAN, J. H. and VAN GENDEREN, H. (1972) Tissue levels in animals and effects caused by chlorinated hydrocarbons, chlorinated biphenyls and mercury in the marine environment. In *Marine Pollution and Sea Life*, pp. 1–8. West Byfleet: Fishing New Books.

KOSS, G., SCHULER, E., ARNDT, B., SIEDEL, J., SEUBERT, S. and SEUBERT, A. (1986) A comparative toxicological study on pike (*Esox lucius* L.) from the River Rhine and River Lahn. *Aquatic Toxicology* **8**, 1–9.

KRUCKEBERG, A. L. and WU, L. (1992) Copper tolerance and copper accumulation of herbaceous plants colonizing inactive Californian copper mines. *Ecotoxicology and Environmental Safety* **23**, 307–319.

KUDO, A., MIYAHARA, S. and MILLER, D. R. (1980) Movement of mercury from Minimata Bay into Yatsushiro Sea. *Progress in Water Technology* **12**, 509–524.

KUHR, R. and DOROUGH, W. (1977) *Carbamate Insecticides*. Cleveland, OH: CRC Press.

LACERDA, L. D., PFEIFFER, W. C., OTT, A. T. and SILVEIRA, E. G. (1989) Mercury contamination in the Madeira River, Amazon – Hg inputs to the environment. *Biotropica* **21**, 91–93.

LEAHY, J. P. (1985) *The Pyrethroid Insecticides*. London: Taylor and Francis.

LEIVESTAD, H., HENDRY, G., MUNIZ, I. P. and SNEKVIK, E. (1976) Effects of acid precipitation on freshwater organisms. In BRAKKE, F. H. (ed.) *Impact of Acid Precipitation on Forest and Freshwater Ecosystems in Norway*, pp. 87–111. Res. Rep. FR 6/76. SNSF Project, Oslo, Norway.

LEPP, N. W. and DICKINSON, N. M. (1994) Fungicide-derived copper in tropical plantation crops. In ROSS, S. M. (ed.) *Toxic Metals in Soil–Plant Systems*, pp. 367–393. Chichester: John Wiley.

LEWIS, M. A. (1990) Are laboratory-derived toxicity data for freshwater algae worth the effort? *Environmental Toxicology and Chemistry* **9**, 1279–1284.

LIMA, S. L. and DILL, L. M. (1990) Behavioural decisions made under the risk of predation: a review and prospectus. *Canadian Journal of Zoology* **68**, 619–640.

LITTLE, E. J., McCAFFERY, A. R., WALKER, C. H. and PARKER, T. (1989) Evidence for an enhanced metabolism of cypermethrin by a monooxygenase in a pyrethroid-resistant strain of the tobacco budworm (*Heliothis virescens* F.). *Pesticide Biochemistry and Physiology* **34**, 58–68.

LOGANATHAN, B. G., TANABE, S., TANAKA, H., WATANABE, S., MIYAZAKI, N., AMANO, M. and TATSUKAWA, R. (1990) Comparison of organochlorine residue levels in the striped dolphin from Western Northern Pacific, 1978–1979 and 1986. *Marine Pollution Bulletin* **21**, 435–439.

LOVELOCK, J. (1989) *The Ages of Gaia: a Biography of our Living Planet*. Oxford: Oxford University Press.

LOWE, V. P. W. (1991) Radionuclides and the birds at Ravenglass. *Environmental Pollution* **70**, 1–26.

LUKASHEV, V. K. (1993) Some geochemical and environmental aspects of the Chernobyl nuclear accident. *Applied Geochemistry* **8**, 419–436.

LUTGENS, F. K. and TARBUCK, E. J. (1992) *The Atmosphere: an Introduction to Meteorology*. Hemel Hempstead: Prentice Hall.

MACKENZIE, A. B. and SCOTT, R. D. (1993) Sellafield waste radionuclides in Irish Sea intertidal and salt marsh sediments. *Environmental Geochemistry and Health* **15**, 173–184.

MACKENZIE, A. B., SCOTT, R. D., ALLAN, R. L., BEN SHABAN, Y. A., COOK, G. T. and PULFORD, I. D. (1994) Sediment radionuclide profiles: implications for mechanisms of Sellafield waste dispersal in the Irish Sea. *Journal of Environmental Radioactivity* **23**, 36–69.

MACNAIR, M. R. (1981) Tolerance of higher plants to toxic materials. In BISHOP, J. A. and COOK, L. M. (eds) *Genetic Consequences of Man-made Change*. London: Academic Press.

MACNAIR, M. R. (1987) Heavy metal tolerance in plants: a model evolutionary system. *Trends in Evolution and Ecology* **2**, 354–359.

MALINS, D. C. and COLLIER, T. K. (1981) Xenobiotic interactions in aquatic organisms: effects on biological systems. *Aquatic Toxicology* **1**, 257–268.

MALLET, J. (1989) The evolution of insecticide resistance: have the insects won? *Trends in Evolution and Ecology* **4**, 336–340.

MALTBY, L. and NAYLOR, C. (1990) Preliminary observations on the ecological relevance of the *Gammarus* 'scope for growth' assay: effect of zinc on reproduction. *Functional Ecology* **4**, 393–397.

MALTBY, L., NAYLOR, C. and CALOW, P. (1990a) Effect of stress on a freshwater benthic detrivore: scope for growth in *Gammarus pulex*. *Ecotoxicology and Environmental Safety* **19**, 285–291.

MALTBY, L., NAYLOR, C. and CALOW, P. (1990b) Field deployment of a scope for growth assay *Gammarus pulex* a freshwater benthic detrivore. *Ecotoxicology and Environmental Safety* **19**, 292–300.

MANAHAN, E. E. (1994) *Environmental Chemistry*, 6th edn. Boca Raton, FL: Lewis.

MARKERT, B. and WECKERT, V. (1994) Higher lead concentrations in the environment of former West Germany after the fall of the 'Berlin Wall'. *Science of the Total Environment* **158**, 93–96.

MARSHALL, J. (1992) Weeds. In GREIG-SMITH, P., FRAMPTON, G. and HARDY, T. (eds) *Pesticides, Cereal Farming and the Environment. The Boxworth Project*. London: HMSO.

MARTIN, M. H. and BULLOCK, R. J. (1994) The impact and fate of heavy metals in an oak woodland ecosystem. In ROSS, S. M. (ed.) *Toxic Metals in Soil–Plant Systems*, pp. 327–365. Chichester: John Wiley.

MARTINEAU, D., LAGACÉ, A., BÉLAND, P., HIGGINS, C. R., ARMSTRONG, D. and SHUGART, L. R. (1988) Pathology of stranded beluga whales (*Delphinapterus leucas*) from the St Lawrence estuary (Quebec, Canada). *Journal of Comparative Pathology* **38**, 287–308.

MCCARTHY, J. F. and SHUGART, L. R. (1990) Biological markers of environmental contamination. In MCCARTHY, J. F. and SHUGART, L. R. (eds) *Biomarkers of Environmental Contamination*, pp. 3–14. Boca Raton, FL: Lewis.

MCNEILLY, T. (1968) Evolution in closely adjacent plant populations III *Agrostis tenvis* on a small copper mine. *Heredity* **23**, 99–108.

MEHLHORN, H., O'SHEA, J. M. and WELLBURN, A. R. (1991) Atmospheric ozone interacts with stress ethylene formation by plants to cause visible plant injury. *Journal of Experimental Botany* **42**, 17–24.

References

MELLANBY, K. (1967) *Pesticides and Pollution.* London: Collins.
MERIAN, E. (ed.) (1991) *Metals and their Compounds in the Environment.* Weinheim: VCH.
MINEAU, P. (ed.) (1991) *Cholinesterase – Inhibiting Insecticides: Their impact on Wildlife and the Environment.* Amsterdam: Elsevier.
MINEAU, P. and PEAKALL, D. B. (1987) An evaluation of avian impact assessment techniques following broad-scale forest insecticide spraying. *Environmental Toxicology and Chemistry* 6, 781–791.
MINEAU, P., FOX, G. A., NORSTROM, R. J., WESELOH, D. V., HALLETT, D. J. and ELLENTON, J. A. (1984) Using the herring gull to monitor levels and effects of organochlorine contamination in the Canadian Great lakes. *Advances in Environmental Science and Technology* 14, 425–452.
MOORE, N. W. and WALKER, C. H. (1964) Organic chlorine insecticide residues in wild birds. *Nature* 201, 1072–1073.
MORAN, P. J. and GRANT, T. R. (1991) Transference of marine fouling communities between polluted and unpolluted sites: impact on structure. *Environmental Pollution* 72, 89–102.
MORIARTY, F. (ed.) (1975) *Organochlorine Insecticides: Persistent Organic Pollutants.* London: Academic Press.
MORIARTY, F. (1988) *Ecotoxicology*, 2nd edn. London: Academic Press.
MORIARTY, F. and WALKER, C. H. (1987) Bioaccumulation in food chains – a rational approach. *Ecotoxicology and Environmental Safety* 13, 208–215.
MOSE, D. G., MUSHRUSH, G. W. and CHROSNIAK, C. E. (1992) A two-year study of seasonal indoor radon variations in southern Maryland. *Environmental Pollution* 76, 195–199.
NAYAK, B. N. and PETRAS, M. L. (1985) Environmental monitoring for genotoxicity: *in vivo* sister chromatid exchange in the house mouse (*Mus musculus*). *Canadian Journal of Genetics and Cytology* 27, 351–356.
NAYLOR, C., MALTBY, L. and CALOW, P. (1989) Scope for growth in *Gammarus pulex*, a freshwater benthic detrivore. *Hydrobiologia*, 188/189, 517–523.
NEBERT, D. W. and GONZALEZ, F. J. (1987) P450 Genes: structure, evolution and regulation. *Annual Review of Biochemistry* 56, 945–993.
NELSON, R. W. (1976) Behavioural aspects of egg breakage in peregrine falcons. *Canadian Field-Naturalist* 90, 320–329.
NEWTON, I. (1986) *The Sparrowhawk.* Calton: Poyser.
NEWTON, I. (1988) Determination of critical pollutant levels in wild populations, with examples from organochlorine insecticides in birds of prey. *Environmental Pollution* 55, 29–40.
NEWTON, I. and HAAS, M. B. (1984) The return of the sparrowhawk. *British Birds* 77, 47–70.
NEWTON, I. and WYLLIE, I. (1992) Recovery of a sparrowhawk population in relation to declining pesticide contamination. *Journal of Applied Ecology* 29, 476–484.
NIEBOER, E. and RICHARDSON, D. H. S. (1980) The replacement of the nondescript term 'heavy metals' by a biologically and chemically significant classfication of metal ions. *Environmental Pollution* 1B, 3–26.
NRC (NATIONAL RESEARCH COUNCIL) (1985) *Oil in the Sea: Inputs Fates and Effects.* Washington, DC: National Academy Press.
OECD (1991) *The State of the Environment.* Paris: OECD.
OPPENOORTH, F. J. (1985) Biochemistry and genetics of insecticide resistance. In

KERKUT, G. A. and GILBERT, L. I. (eds) *Comprehensive Insect Physiology Bio-chemistry and Pharmacology*, 12. Oxford: Pergamon Press.

PAASIVIRTA, J. (1991) *Chemical Ecotoxicology*. Chelsea, MI: Lewis.

PAYNE, J. F., FANCEY, L. L., RAHIMTULA, A. D. and PORTER, E. L. (1987) Review and perspective on the use of mixed-function oxygenase enzymes in biological monitoring. *Comparative Pharmacology and Physiology* **86C**, 233–245.

PEAKALL, D. B. (1985) Behavioural responses of birds to pesticides and other contaminants. *Residue Reviews* **96**, 45–77.

PEAKALL, D. B. (1992) *Animal Biomarkers as Pollution Indicators*. London: Chapman and Hall.

PEAKALL, D. B. (1993) DDE-induced eggshell thinning: an environmental detective story. *Environmental Reviews* **1**, 13–20.

PEAKALL, D. B. and BART, J. R. (1983) Impacts of aerial application of insecticides on forest birds. *Critical Reviews in Environmental Control* **13**, 117–165.

PEAKALL, D. B. and SHUGART, L. R. (eds) (1993) *Biomarkers. Research and Application in the Assessment of Environmental Health*. Berlin: Springer-Verlag.

PEARCE, P. A., PEAKALL, D. B. and ERSKINE, A. J. (1976) Impact on forest birds of the 1975 spruce budworm spray operations in New Brunswick. CWS Progress Notes No. 62.

PERKINS, D. F. and MILLAR, R. O. (1987) Effects of airborne fluoride emissions near an aluminium works in Wales: Part 1 – corticolous lichens growing on broad-leaved trees. *Environmental Pollution* **47**, 63–78.

PFEIFFER, W. C., DE LACERDA, L. D., MALM, O., SOUZA, C. M. M., DA SILVEIRA, E. G. and BASTOS, W. R. (1989) Mercury concentrations in inland waters of gold-mining areas in Rondonia, Brazil. *Science of the Total Environment* **87/88**, 233–240.

PFEIFFER, W. C., DE LARCERDA, L. D., SALOMONS, W. and MALM, O. (1993) Environmental fate of mercury from gold mining in the Brazilian Amazon. *Environmental Research* **1**, 26–37.

PICKERING, Q. C., HENDERSON, C. and LEMKE, A. E. (1962) The toxicity of organic phosphorus insecticides to different species of warmwater fishes. *Transactions of the American Fisheries Society* **91**, 175–184.

PILLING, E. D., BROMLEY-CHALLENOR, K. A. C., WALKER, C. H. and JEPSON, P. C. (1995) Mechanisms of synergism between the pyrethroid insecticide *t*-cyhalothrin and the imidazole fungicide prochloraz in the honeybee (*Apis mellifera*). *Pesticide Biochemistry and Physiology* **51**, 1–11.

PLENDERLEITH, R. W. and BELL, L. C. (1990) Tolerance of twelve tropical grasses to high soil concentrations of copper. *Tropical Grasslands* **24**, 103–110.

PORTEOUS, A. (1992) *Dictionary of Environmental Science and Technology*. Chichester: John Wiley.

POSTEL, S. (1984) Air Pollution, Acid Rain and the Future of Forests. Worldwatch Paper 58.

POSTHUMA, L. and VAN STRAALEN, N. M. (1993) Heavy-metal adaptation in terrestrial invertebrates: a review of occurrence, genetics, physiology and ecological consequences. *Comparative Biochemistry and Physiology* **106C**, 11–38.

POTTS, G. R. (1986) *The Partridge*. London: Collins.

POTTS, G. R. (in press) The grey partridge. In PAIN, D. and DIXON, J. (eds) *Bird Conservation and Farming Policy in the European Union*. London: Academic Press.

PREIN, A. E., THIE, G. M., ALINK, G. M., KOEMAN, J. H. and POELS,

C. L. M. (1978) Cytogenic changes in fish exposed to water of the River Rhine. *Science of the Total Environment* **9**, 287–291.

PURDOM, C. E., HARDIMAN, P. A., BYE, V. J., ENO, N. C., TYLER, C. R. and SUMPTER, J. P. (1994) Estrogenic effects of effluents from sewage treatment works. *Chemistry Ecology* **8**, 275–285.

RAMADE, F. (1992)*Précis d'Ecotoxicologie*. Paris: Masson.

RATCLIFFE, D. A. (1967) Decrease in eggshell weight in certain birds of prey. *Nature* **215**, 208–210.

RATCLIFFE, D. (1993) *The Peregrine Falcon*, 2nd edn. Calton: T and D Poyser.

RICHTER, C. A., DRAKE, J. B., GIESY, J. P. and HARRISON, R. O. (1994) Immunoassay monitoring of polychlorinated biphenyls (PCBs) in the Great Lakes. *Environmental Science and Pollution Research* **1**, 69–74.

RIDLEY, M. (1993) *Evolution*. Oxford: Blackwell Scientific Publications.

RISEBROUGH, R. W. and PEAKALL, D. B. (1988) The relative importance of the several organochlorines in the decline of Peregrine Falcon populations. In CADE, T. J., ENDERSON, J. H., THELANDER, C. G. and WHITE, C. M. (eds) *Peregrine Falcon Populations: their Management and Recovery*, pp. 449–468. Boise, ID: The Peregrine Fund, Inc.

ROBINSON, J., RICHARDSON, A., CRABTREE, A. N., COULSON, J. C. and POTTS, G. R. (1967) Organochlorine residues in marine organisms. *Nature* **214**, 1307–1311.

RONIS, M. J. J. and WALKER, C. H. (1989) The microsomal monooxygenases of birds. *Reviews in Biochemical Toxicology* **10**, 310–384.

ROSMAN, K. J. R., CHISHOLM, W., BOUTRON, C. F., CANDLESTONE, J. P. and GÖRLACH, U. (1993) Isotopic evidence for the source of lead in Greenland snows since the late 1960s. *Nature* **362**, 333–335.

ROSS, S. M. (ed.) (1994) *Toxic Metals in Soil–Plant Systems*. Chichester: John Wiley.

ROUSH, R. T. and DALY, D. C. (1990) The role of population genetics in resistance research and management. In ROUSH, R. T. and TABASHNIK, B. E. (eds) *Pesticide Resistance in Arthropods*. New York: Chapman and Hall.

SAFE, S. (1990) Polychlorinated biphenyls (PCBs), dibenzo-p-dioxins (PCDDs), dibenzofurans (PCDFs), and related compounds: Environmental and mechanistic considerations which support the development of toxic equivalency factors (TEFs). *Critical Reviews in Toxicology* **21**, 51–88.

SALOMONS, W., BAYNE, B. L., DUURSMA, E. K. and FÜRSTNER, V. (eds) (1988) *Pollution of the North Sea: an Assessment*. Berlin: Springer-Verlag.

SANDERS, B. M. (1993) The cellular stress response. *Environs* **16**, 3–6.

SAWICKI, R. M. and RICE, A. D. (1978) Response of susceptible and resistant peach-potato aphid *Myzus persicae* (Sulz.) to insecticides in leaf-dip assays. *Pesticide Science* **9**, 513–516.

SCHAT, H. and BOOKUM, W. M. T. (1992) Genetic control of copper tolerance in *Silene vulgaris, Heredity* **68**, 219–229.

SCHEUHAMMER, A. M. (1989) Monitoring wild bird populations for lead exposure. *Journal of Wildlife Management* **53**, 759–765.

SCHINDLER, D. W. (in press) Ecosystems and ecotoxicology: a personal perspective. In NEWMAN, M. and JAGOE, C. (eds) *Quantitative Ecotoxicology: a Hierarchical Approach*. Chelsea, MI: Lewis.

SCHINDLER, D. W., FROST, T. M., MILLS, K. H., CHANG, P. S. S., DAVIES, I. J., FINDLAY, L., MALLEY, D. F., SHEARER, J. A., TURNER, M. A.,

Principles of ecotoxicology

GARRISON, P. J., WATRAS, C. J., WEBSTER, K., GUNN, J. M., BREZONIK, P. L. and SWENSON, W. A. (1991) Comparisons between experimentally and atmospherically-acidified lakes during stress and recovery. *Proceedings of the Royal Society of Edinburgh* **97B**, 193–226.

SCHMITT, C. J. and BRUMBAUGH, W. G. (1990) National contaminant biomonitoring program: concentrations of arsenic, cadmium, copper, lead, mercury, selenium and zinc in US freshwater fish, 1976–1984. *Archives of Environmental Contamination and Toxicology* **19**, 731–747.

SEAGER, J. and MALTBY, L. (1989) Assessing the impact of episodic pollution. *Hydrobiologia* **188/189**, 633–640.

SHAW, A. J. (ed.) (1989) *Heavy Metal Tolerance in Plants: Evolutionary Aspects.* Boca Raton, FL: CRC Press.

SHUGART, L. (1994) Genotoxic responses in blood. In FOSSI, M. C. and LEONZIO, C. (eds) *Nondestructive Biomarkers in Vertebrates*, pp. 131–145. Boca Raton, FL: Lewis.

SIBLY, R. M. (1996) Effects of pollutants on individual life histories and population growth rates. In NEWMAN, M. C. and JAGOE, C. (eds) *Quantitative Ecotoxicology: a Hierarchical Approach.* Chelsea, MI: Lewis.

SIBLY, R. M. and ANTONOVICS, J. (1992) Life-history evolution. In BERRY, R. J., CRAWFORD, T. J. and HEWITT, G. M. (eds) *Genes in Ecology.* Oxford: Blackwell Scientific Publications.

SIBLY, R. M. and CALOW, P. (1989) A life cycle theory of responses to stress. *Biological Journal of the Linnean Society* **37**, 110–116.

SIBUET, M., CALMET, D. and AUFFRET, G. (1985) Reconnaissance photographique de conteneurs en place dans la zone d'immersion des dechets faiblement radioactifs de l'Atlantique Nord-Est. *Compte Rendu Hebdomadaire des Seances de l'Academie des Sciences* Paris, Serie III, **301**, 497–502.

SIMKISS, K. (1993) Radiocaesium in natural systems – a UK coordinated study. *Journal of Environmental Radioactivity* **18**, 133–149.

SINCLAIR, A. R. E. (1989) Population regulation in animals. In CHERRETT, J. M. (ed.) *Ecological Concepts.* Oxford: Blackwell Scientific Publications.

SÖDERGREN, A. (1991) Environmental fate and effects of bleached pulp mill effluents. Swedish Environment Protection Agency Report 4031.

SOLOMONS, W. G. (1986) *Organic Chemistry*, 2nd edn. New York: Wiley.

SOLONEN, T. and LODENIUS, M. (1990) Feathers of birds of prey as indicators of mercury contamination in Southern Finland. *Holarctic Ecology* **13**, 229–237.

SOMERVILLE, L. and GREAVES, M. P. (1987) *Pesticide effects on Soil Microflora.* London: Taylor and Francis.

SOMERVILLE, L. and WALKER, C. H. (eds) (1990) *Pesticide Effects on Terrestrial Wildlife.* London: Taylor and Francis.

SPEAR, P. A., MOON, T. W. and PEAKALL, D. B. (1986) Liver retinoid concentrations in natural populations of herring gulls (*Larus argentatus*) contaminated by 2,3,7,8-tetrachlorodibenzo-p-dioxin and in ring doves (*Streptopelia risoria*) injected with a dioxin analogue. *Canadian Journal of Zoology* **64**, 204–208.

SPURGEON, D. J., HOPKIN, S. P. and JONES, D. T. (1994) Effects of cadmium, copper, lead and zinc on growth, reproduction and survival of the earthworm *Eisenia fetida* (Savigny): assessing the environmental impact of point-source metal contamination in terrestrial ecosystems. *Environmental Pollution* **84**, 123–130.

STEIN, J. E., COLLIER, T. K., REICHERT, W. L., CASILLAS, E., HOM, T. and

312

VARANASI, U. (1992) Bioindicators of contaminant exposure and sublethal effects: studies with benthic fish in Puget Sound, Washington. *Environmental Toxicology and Chemistry* **11**, 701–714.

TARDIFF, R. (ed.) (1992) *Methods to Assess Adverse Effects of Pesticides on Non-target Organisms. Scope 49.* Chichester: John Wiley.

TAYLOR, C. E. (1986) Genetics and evolution of resistance to insecticides. *Biological Journal of the Linnean Society* **27**, 103–112.

TERRIERE, L. C. (1984) Induction of detoxification enzymes in insects. *Annual Review of Entomology* **29**, 71–88.

TIMBRELL, J. A. (1992) *Principles of Biochemical Toxicology*, 2nd edn. London: Taylor and Francis, 416 pp.

TIMBRELL, J. A. (1995) *Introduction to Toxicology*, 2nd edn. London: Taylor and Francis, 168 pp.

TOLBA, M. K. (1992) *Saving our Planet.* London: Chapman and Hall, 287 pp.

TUREKIAN, K. K. (1976) *Oceans*, 2nd edn. Hemel Hempstead: Prentice-Hall.

TURLE, R., NORSTROM, R. J. and COLLINS, B. (1991) Comparison of PCB quantitation methods: re-analysis of archived specimens of herring gull eggs from the Great Lakes. *Chemosphere* **22**, 201–213.

TURTLE, E. E., TAYLOR, A., WRIGHT, E. N., THEARLE, R. J. P., EGAN, H. EVANS, W. H. and SOUTAR, N. M. (1963) The effects on birds of certain chlorinated insecticides used as seed dressings. *Journal of Science in Food and Agriculture* **8**, 567–576.

UNEP (1993) *Environmental Data Report 1993–94.* Oxford: Blackwell.

VAN DER GAAG, M. A., VAN DER KERKHOF, J. F. J., VAN DER KLIFT, H. W. and POELS, C. L. M. (1983) Toxicological assessment of river quality in bioassays with fish. *Environmental Monitoring and Assessment* **3**, 247–255.

VAN STRAALEN, N. M. (1993) Soil and sediment quality criteria derived from invertebrate toxicity data. In DALLINGER, R. and RAINBOW, P. S. (eds) *Ecotoxicology of Metals in Invertebrates*, pp. 427–441. Chelsea, USA: Lewis.

VARANASI, U., REICHART, W. L. and STEIN, J. E. (1989) 32P-Postlabelling analysis of DNA adducts in liver of wild english sole (*Parophys vetulus*) and winter flounder (*Pseudopleuronectes americanus*). *Cancer Research* **49**, 1171–1177.

VASQUEZ, M. D., POSCHENRIEDER, C., BARCELO, J., BAKER, A. J. M., HATTON, P. and COPE, G. H. (1994) Compartmentation of zinc in roots and leaves of the zinc hyperaccumulator *Thlaspi caerulescens* J and C Presl. *Botanica Acta* **107**, 243–250.

WAID, J. S. (ed.) (1985–7) *PCBs in the Environment*, vols I–III. Cleveland, USA: CRC Press.

WALKER, C. H. (1975) *Environmental Pollution by Chemicals*, 2nd edn. London: Hutchinson.

WALKER, C. H. (1980) Species differences in some hepatic microsomal enzymes that metabolize xenobiotics. *Progress in Drug Metabolism* **5**, 118–164.

WALKER, C. H. (1983) Pesticides and birds: mechanisms of selective toxicity. *Agricultural Ecosystems and Environ* **9**, 211–226.

WALKER, C. H. (1987) Kinetic models for predicting bioaccumulation of pollutants in ecosystems. *Environmental Pollution* **44**, 227–240.

WALKER, C. H. (1990a) Persistent pollutants in fish-eating sea birds: bioaccumulation metabolism and effects. *Aquatic Toxicology* **17**, 293–324.

WALKER, C. H. (1990b) Kinetic models to predict bioaccumulation of pollutants.

Functional Ecology **4**, 295–301.

WALKER, C. H. (1992) Biochemical responses as indicators of toxic effects of chemicals in ecosystems. *Toxicology Letters* **64/65**, 527–532.

WALKER, C. H. (1994) Comparative toxicology (Chapter 9). In LEVI, P. and HODGSON, E. (eds) *Introduction to Biochemical Toxicology*, pp. 193–218. Norwalk, CT: Appleton and Lange.

WALKER, C. H. and LIVINGSTONE D. R. (1992) *Persistent Pollutants in Marine Ecosystems*. Oxford: Pergamon Press, 272 pp.

WALKER, C. H. and THOMPSON, H. M. (1991) Phylogenetic distribution of cholinesterases and related esterases. In MINEAU, P. (ed.) *Cholinesterase-Inhibiting Insecticides: Impact on Wildlife and the Environment*, pp. 1–17. Amsterdam: Elsevier.

WALKER, C. H., MACKNESS, M. I., BREARLEY, C. J. and JOHNSTON, G. O. (1991a) Toxicity of pesticides to birds; the enzymic factor. *Biochemical Society Transactions* **19**, 741–745.

WALKER, C. H., GREIG-SMITH, P. W., CROSSLAND, N. O. and BROWN, R. (1991b) Ecotoxicology. In BALLS, M., BRIDGES, J. and SOUTHEE, J. (eds) *Proceedings of FRAME meeting 'Animals and Alternatives in Toxicology – Present Status and Future Prospects'*. London: MacMillan, pp. 223–252.

WALKER, C. H., JOHNSTON, G. O. and DAWSON, A. (1993) Enhancement of toxicity due to the interaction of pollutants at the toxicokinetic level in birds. *Science of the Total Environment* **134(s)**, 525–531.

WALTON, K. C. (1986) Fluoride in moles, shrews and earthworms near an aluminium reduction plant. *Environmental Pollution* **42A**, 361–371.

WARD, D. V., HOWES, B. L. and LUDWIG, D. F. (1976) Interactive effects of predation pressure and insecticide (Temefos) toxicity on populations of the marsh fidder crab *Uca pugnax*. *Marine Biology* **35**, 119–126.

WARREN, C. E. and DAVIS, G. L. (1967) Laboratory studies on the feeding of fishes. In GERKING, S. D. (ed.) *The Biological Basis of Freshwater Fish Production*, pp. 175–214. Oxford: Blackwell Scientific Publications.

WATERMAN, M. R. and JOHNSON, E. F. (eds) (1991) *Methods in Enzymology*, **206**. San Diego: Academic Press.

WATT, W. D., SCOTT, C. D. and WHITE, W. J. (1983) Evidence of acidification of some Novia Scotia rivers and its impact on Atlantic salmon, *Salmo salar*. *Canadian Journal of Fisheries and Aquatic Science* **40**, 462–473.

WAYNE, R. P. (1991) *Chemistry of Atmospheres*, 2nd edn. Oxford: Clarendon Press.

WESELOH, D. V., TEEPLE, S. M. and GILBERTSON, M. (1983) Double-crested cormorants of the Great Lakes: egg-laying parameters, reproductive failure and contaminant residues in eggs, Lake Huron 1972–1973. *Canadian Journal of Zoology* **61**, 427–436.

WIDDOWS, J. and DONKIN, P. (1991) Role of physiological energetics in ecotoxicology. *Comparative Biochemistry and Physiology* **100C**, 69–75.

WIDDOWS, J. and DONKIN, P. (1992) Mussels and environmental contaminants: bioaccumulation and physiological aspects. In GOSLING, E. (ed.) *The Mussel Mytilus: Ecology, Physiology, Genetics and Culture*. Amsterdam: Elsevier.

WIDDOWS, J., BURNS, K. A., MENON, K. R., PAGE, D. S. and SORIA, S. (1990) Measurement of physiological energetics (scope for growth) and chemical contaminants in mussels (*Arca zebra*) transplanted along a contamination gradient in Bermuda. *Journal of Experimental Marine Biology and Ecology* **138**, 99–117.

WIGFIELD, D. C., WRIGHT, S. C., CHAKRABARTI, C. L. and KARWOWSKA, R. (1986) Evaluation of the relationship between chemical and biological monitoring of low level lead poisoning. *Journal of Applied Toxicology* **6**, 231–235.

WILKINSON, C. F. (1976) Insecticide interactions (Chapter 15). In WILKINSON, C. F. (ed.) *Insecticide Biochemistry and Physiology*. New York: Plenum Press.

WILLIAMS, R. J. P. (1981) Natural selection of the chemical elements. *Proceedings of the Royal Society of London* **213B**, 361–397.

WOOD, M. (1995) *Environmental Soil Biology*. Glasgow: Blackie.

WOOD, R. J. and BISHOP, J. A. (1981) Insecticide resistance: populations and evolution. In BISHOP, J. A. and COOK L. M. (eds) *Genetic Consequences of Man-made Change*, pp. 97–127. London: Academic Press.

WONG, S., FOURNEIR, M., CODERRE, D., BANSKA, M. and KRZYSTYNIAK, K. (1992) Environmental immunotoxicology. In PEAKALL, D. B. (ed.) *Animal Biomarkers as Pollution Indicators*, pp. 166–189. London: Chapman and Hall.

WRIGHT, J. F., FURSE, M. T. and ARMITAGE, P. D. (1993) RIVPACS – a technique for evaluating the biological quality of rivers in the UK. *European Water Pollution Control* **3**, 15–25.

YOKOI, K., KIMURA, M. and ITOKAWA, Y. (1990) Effect of dietary tin deficiency on growth and mineral status in rats. *Biological Trace Element Reseach* **24**, 223–231.

Index

317

U501   5-12

13